NATIONAL AUDUBON SOCIETY POCKET GUIDE

CLOUDS AND STORMS

Text by Dr. David M. Ludlum, Ronald L. Holle, and Dr. Richard A. Keen
Alfred A. Knopf, New York

This is a Borzoi Book.
Published by Alfred A. Knopf, Inc.

Copyright © 1995, 1998 by Chanticleer Press, Inc.
All rights reserved under International and Pan-American
Copyright Conventions. Published in the United States
by Alfred A. Knopf, Inc., New York, and simultaneously in
Canada by Random House of Canada Limited, Toronto.
Distributed by Random House, Inc., New York.

Prepared and produced by Chanticleer Press, Inc.,
New York.
Typeset by Chanticleer Press, Inc., New York.
Printed and bound by Toppan Printing Co., Ltd.,
Hong Kong.

Published March 1995
Second Printing, December 1998

Library of Congress Catalog Card Number: 94-42431
ISBN: 0-679-77999-X

National Audubon Society® is a registered trademark of
National Audubon Society, Inc., all rights reserved.

Contents

How to Use This Guide

Weather, an often powerful and pervasive force, affects us in countless ways. Rain, snow, heat, cold, and drought can play dramatic roles in our lives, while the subtle beauty of cloud formations and rainbows has gentler effects. An understanding of weather-related events has practical benefits and can deepen our appreciation of the natural world.

Coverage

Built around 80 color plates, this book is a guide to the fundamental elements of weather, with an emphasis on North America. In addition to extensive coverage of major cloud types and storm systems, it explains optical phenomena such as rainbows and auroras.

Organization

This easy-to-use guide is divided into three parts: an introduction, an illustrated description of cloud and storm systems, and an appendix.

Introduction

The essay "What Is Weather?" provides an overview of how weather originates. It explains how large-scale airflow patterns are driven into motion by the Sun's heat and the Earth's rotation. Highs, lows, warm fronts, and cold fronts are described in simple language, as is the movement of storms across the continent. The essay concludes with discussion of how clouds and precipitation form. "Cloud Types" describes the different categories of clouds. The Key to Clouds and Storms identifies the margin drawings that accompany the text descriptions.

The Clouds and Storms	This section starts with satellite views of atmospheric events on Earth. These are followed by photographs of more than a dozen major cloud types. The next section depicts the conditions at sky and ground level during major weather events: thunderstorms, tornadoes, hurricanes, and snowstorms. The final plates are images of beautiful optical phenomena related to the weather. Each color plate is accompanied by a detailed text description of the cloud, storm, or phenomenon pictured. For quick reference, a black-and-white drawing illustrates the weather phenomenon shown in the photograph.
Appendix	In an easy-to-consult format, a table titled "A Summary of Cloud Types and Their Significance" explains what the major cloud types tell us about impending weather.

This guide will teach you not only to observe, but also to understand and interpret what you see, deepening your appreciation of daily and seasonal weather conditions.

What Is Weather?

Weather is a common ingredient of our everyday lives, yet for most people it is more mysterious than many other aspects of nature. Knowledge of the forces and phenomena involved in the weather can remove some of the mystery and allow the average person to understand and interpret meteorological events. Technically speaking, weather is the instantaneous and constantly changing state of the atmosphere. It is described on the global scale by maps that show observations of temperature, pressure, wind, moisture, and other data at selected heights above the ground. On the local scale, weather may be described by the current temperature, pressure, wind, and other readings, along with what are called the *sky conditions:* the types, amounts, and heights of clouds, plus notes on whatever else might be in the air, such as rain, snow, fog, dust, smoke, and other visible phenomena (like rainbows). This guide is a compact catalog of the visible sky phenomena that relate to Earth's weather.

Large-Scale Airflow Patterns

Earth's atmosphere is driven into motion by the uneven heating of our planet by the Sun. In the tropics, where the noontime Sun passes nearly overhead, the ground and the air above it receive much more solar heat than in polar regions, where the Sun is at lower angles and therefore spreads the available sunlight over much larger areas. Heated tropical air rises and spreads outward toward the north and south poles, while cold polar air sinks and flows toward the equator in lower levels of the atmosphere. This airflow is called the *Hadley circulation,* after the Englishman who proposed the concept in 1735.

This theoretically simple up-and-down, north-to-south Hadley circulation is warped into the complex pattern of winds we actually see by the *Coriolis effect,* which is the result of Earth's rotation. While our 7,927-mile-diameter (12,756-km-diameter) Earth rotates once every 24 hours, everything on the equator (including air) moves to the east at 1,038 mph (1,670 kph). This movement decreases to 0 mph at the north and south poles. As the Hadley circulation rises and leaves the vicinity of the equator, its eastward motion of 1,038 mph becomes increasingly greater than that of the ground below. By the time the Hadley circulation reaches 20° or 30° N (and S) latitude, the speed difference becomes so great that the air no longer flows toward the poles but is deflected eastward into the subtropical jet stream, one of the strongest and most consistent winds of the world. The Coriolis effect deflects all moving air on Earth to some degree and is responsible for the spiral motions found in hurricanes, tornadoes, and other storms.

The patterns of uneven heating and circulating air currents change during the course of a year due to the effect of the seasons. On the first day of Northern Hemisphere summer the noontime Sun passes directly overhead at 23.4° N latitude, while on the first day of winter the Sun's noontime overhead passage is at 23.4° S. Accordingly, the latitude of greatest solar heating is farther north during the summer months, and atmospheric currents in turn shift to the north. Meanwhile, higher Sun angles and longer days in the Arctic during summer decrease the heat difference between the tropics and the North Pole and weaken the heat-driven currents. Thus Northern Hemisphere weather patterns are, in general, farther north and weaker during the summer.

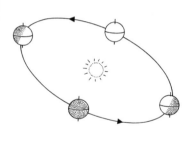

Most of Earth between 30° and 60° N (and S) latitude lies in the zone of the *prevailing westerlies* (winds blowing from the west). This band includes most of the U.S. and southern Canada, but not the Florida peninsula, Mexico, and Hawaii (which lie under the trade-wind easterlies much of the year), and parts of Alaska and northern Canada (which lie under Arctic easterlies, especially in winter).

The prevailing westerlies may be thought of as a broad stream of air that circles the globe at middle latitudes. During summer this stream moves north into Canada; in winter it shifts south to the middle of the U.S. At all times of year it may stray far from its normal position and meander into bends and bows thousands of miles long.

Highs and Lows Embedded in the broad stream of the prevailing westerlies, smaller whirls and eddies that are responsible for most of our weather move from west to east in never-ending succession. These whirls are the low- and high-pressure systems, also known (respectively) as *cyclones* and *anticyclones,* or simply *lows* and *highs,* that bring alternating spells of foul and fair weather. The word "cyclone" derives from the Greek *kyklos,* meaning "circle"; air in a cyclone blows in a circular pattern around the center. In the Northern Hemisphere, cyclones turn counterclockwise (as seen from above), while in anticyclones the flow is clockwise. These directions are reversed in the Southern Hemisphere. Cyclones are

11

centers of low barometric pressure, and anticyclones have high pressure (in both hemispheres). Low pressure is essentially a partial vacuum, so air streams into cyclones from surrounding areas of higher pressure (anticyclones). Putting all these motions together, air spirals outward from the center of a high-pressure anticyclone and inward toward the center of a low-pressure cyclone.

Cyclones come in all sizes, from tiny whirlwinds like dust devils (miniature low-pressure systems) to huge regions of low pressure covering half a continent. Meteorologists commonly use the term *cyclone* to refer to a storm system that is several hundred to a thousand miles across.

Cyclones nearly always form along the boundary, known as a *front,* between arctic air to the north and subtropical air to the south. There are four basic kinds of fronts. Before a cyclone develops, the boundary between opposing air masses may lie still for a while. These *stationary fronts* are usually fairly tranquil places with slow-moving patches of light rain or snow resulting in protracted spells of drizzly weather. However, if the air on the warm side of the front is exceptionally moist, then heavy, slow-moving thunderstorms may develop, unleashing torrential rains and flash floods.

As the cyclone starts spinning, the stationary front starts moving. A *warm front* marks the leading edge of northbound warm air, usually on the east side of the storm center, while west of the center arctic air plunges south behind a *cold front*. Along cold fronts, heavier cold air shoves like a wedge beneath the warm, and usually moist, air. Forced upward, the warm air expands and cools, and its moisture condenses into clouds, rain, and snow. The forced uplift is often great enough to produce heavy showers and thunderstorms. Along warm fronts the warm air overruns the cold air to its north, with a gentler rising motion producing more moderate, but longer-lasting, rain- and snowfalls.

Both warm and cold fronts rotate like spokes around the low-pressure center, and after a day or two the (usually) faster-moving cold front may catch up to the warm front. The *warm sector*, a wedge of warm air caught between the two fronts, shrinks. Eventually the fronts merge into an *occluded front*, an indistinct boundary between two similarly cold air masses. Although the warm sector is now completely lifted off the ground, several thousand feet up rain and snow are still being made.

Although no two storms are ever the same, passing cyclones often bring a characteristic sequence of weather. The warm front ahead (east) of an approaching cyclone frequently brings thickening and lowering clouds followed by steady precipitation in the form of rain or snow. If the cyclone passes to your south and your location remains within the cold air, precipitation and temperatures may remain steady until the low center moves east and skies clear. If the cyclone passes to your north, passage of the warm front brings higher temperatures. After several hours in the warm sector of the cyclone, the center of low pressure passes, followed by the cold front, which may set off brief but heavy precipitation in showers, squalls, and thunderstorms.

The cyclone's departure brings clearing and cooler weather as the anticyclone, or high, ushers in one or several days of fair skies and drier weather. As the anticyclone passes overhead and moves east, winds turn southerly in advance of the next cyclone.

This is the normal sequence of weather accompanying the passage of traveling cyclones and anticyclones in middle latitudes, but it is wise to remember that "normal" in meteorology usually means the average of many unique events. Thus cyclones and anticyclones may approach from any point of the compass, they may stall overhead, or they may suddenly strengthen or weaken when least expected.

Storm Tracks   Every year several hundred cyclones enter, pass across, or develop over North America; on an average day one to five cyclones dot the continent's weather map. The paths of these cyclones are not completely random; many follow preferred storm tracks across and along the continent. At any given locality, one or two cyclones per week pass close enough to affect the weather, which means that every five days, on average, the local weather does a complete swing from clear to rain (or snow) and back to clear, and from cold to warm to cold again. It is not an even cycle, however. Weeks may pass, especially during summer and early fall, without a significant cyclone affecting the local weather. At these times, huge high-pressure systems over the western Atlantic Ocean (the Bermuda High) and the eastern Pacific expand inland, and much of North America sweats in persistent heat. The clockwise circulation around the Bermuda High usually draws enough humid air from the Gulf of Mexico to breed thunderstorms over and east of the Rocky Mountains. But if the oceanic highs stretch far enough inland, or if another hot high-pressure cell develops over the central plains, the flow of moist air from the tropical oceans is cut off, leading to normal dry seasons in the southwestern U.S. and droughts or dry spells elsewhere.

Once or twice a year, on average, North America is struck or threatened by a tropical storm or hurricane. Such a storm is also a cyclone, but its origins are quite different from those of the mid-latitude cyclones we have described. Hurricanes form over tropical oceans, where the temperature contrasts found along warm and cold fronts are virtually nonexistent and are driven by upward motions of air in massive spiral bands of powerful thunderstorms. Most hurricanes move in a westward direction, in contrast to the normal eastward tracks of mid-latitude cyclones.

Condensation and Clouds   The 1 percent of Earth's atmosphere that is not oxygen and nitrogen contains, among other constituents, water vapor. However, the actual percentage of water vapor is quite variable, ranging from less than 0.1 percent in cold, dry arctic air to about 4 percent in humid tropical

16

air masses. This small, variable amount of water vapor in the air is responsible for most of what we call—and see as—weather.

Water vapor is an essentially invisible gas. To become visible as a cloud, or to fall to the ground as precipitation, the molecules of water vapor must condense into liquid water or solid ice. Warm air can hold more water vapor than cold air can; if air containing water vapor is cooled, it eventually reaches a temperature at which it can no longer hold all its supply of vapor. The excess vapor condenses into tiny drops of liquid water, or, if the air is cold enough, the vapor forms ice crystals through a process called *sublimation*.

The temperature at which condensation occurs in a specific mass of air is called the *dew point;* this is a useful measure of the amount of vapor in the air. The higher the dew point, the more moisture there is in the air. Even though air may cool below its dew point, water vapor still needs something—a surface or a particle—to condense on. Air is full of minuscule particles of dust, pollen, sea salt, smoke, and even tiny drops of turpentine blown from trees, with between a thousand and a million of these *condensation nuclei* in every cubic inch of air.

One way air is cooled down to the dew point is by lifting; as air rises, the surrounding pressure decreases, and the air expands and cools. At the altitude at which the temperature falls to the dew point, condensation occurs. Above this *condensation level* sits a cloud. Often air rises along fronts and in cyclones, as described earlier. Sometimes, particularly in summer, solar heating of the ground sends warm "bubbles" of air up to the condensation level. In mountainous regions, air rises by simply blowing uphill (a process known as *orographic lifting*).

Air may also cool in place by radiating heat out to space or by contact with a cold surface, such as snow-covered ground or a cold water surface. When the ground (or plants on the ground) cools below the dew point, vapor condenses into dew or frost; when the air itself reaches the dew point, fog (a cloud at ground level) forms.

The drops of water that condense to make clouds are incredibly small. They average much less than a thousandth of an inch in diameter; it would take 10,000 of them to cover the head of a pin.

Rain and Snow   Cloud droplets grow by absorbing other droplets through a process called *coalescence*. The drops start falling as rain when they become about $\frac{1}{16}$ inch across; it takes about a million cloud droplets to make one average raindrop. Much of the rain that falls in the tropics forms by coalescence. In the cooler and drier air over most of North America, coalescence rarely produces more than light drizzle. Outside the tropics, most rain begins as snow, partly because any cloud big enough to make precipitation is probably tall enough to have its top portion below freezing, and partly because ice crystals grow so much faster than raindrops.

The size and shape of ice crystals are determined by the temperature and moisture content of the air in which they form. In the extremely dry and cold air at high altitudes (or at the surface in the Arctic in winter) the most common ice crystals are minute, hexagonal columns and flat, six-sided wafers called hexagonal plates. These are the forms of ice crystals found in cirrus clouds. Delicate, six-pointed, star-shaped snowflakes form in moister air in clouds at temperatures between 0° and 20°F (−18° to −7°C). At temperatures closer to the freezing point of

32°F (0°C), ice grows into splinter-shaped pieces known as needles. In moderate to heavy snowstorms, individual crystals stick together in large aggregates up to several inches across. If the temperature in the lower atmosphere is warm enough, the falling snow melts into rain.

Convection   The process of condensation not only is a product of rising air currents but also may cause air to rise. As a vapor, water contains a large amount of heat, called *latent heat.* The water gains this heat from its environment when it evaporates, which is why evaporating water cools its surroundings. When it condenses, the heat is released back into its environment, warming the surrounding air, making it "lighter" and encouraging a rising motion.

If the air contains enough moisture, the release of latent heat may cause the cloud itself to grow vertically. As the air within the cloud rises, it is replaced by new moist air drawn into the base of the cloud. This leads to more condensation, latent heat release, and further rising currents. This process is called convection.

We see these rising blobs as the growing cauliflower-like turrets of *cumulus clouds.* If there is enough moisture in

20

the air and other atmospheric conditions are right, the cumulus cloud may develop into a *swelling cumulus*. When the growing cumulus cloud reaches heights of three or four miles (lower in winter), the temperature of the cloud top falls below 32°F (0°C). Now ice crystals, rather than liquid water droplets, result from condensation. The top of the cumulus cloud may assume the fuzzy and fibrous appearance characteristic of ice clouds. High winds aloft may whip ice crystals from the main cloud into a flat, spreading *anvil* cloud.

The appearance of ice crystals changes the nature of the growing cumulus cloud. With snow crystals growing in the cloud top, precipitation forms much more rapidly and rain begins falling. Now we call the cloud a *cumulonimbus* (the Latin word *nimbus* means rain). In a mature cumulonimbus cloud, updrafts of fresh, moist air coexist with downdrafts full of rain-laden air. Eventually the supply of fresh air either runs out or is cut off by downdraft air spreading outward beneath the cloud, and the cumulonimbus begins to dissipate. The entire life cycle of a cumulonimbus cloud may span one or several hours.

Cloud Types

There are two general types of clouds—cumulus (heaped) and stratus (layered)—which sometimes coexist. Clouds are also named according to their elevation and whether they contain enough moisture to form precipitation.

Cumulus Types

Cumulus clouds, which are rounded on top and may become quite tall, result from convection in unstable air. They may be created by strong heating of the ground, by lifting in the vicinity of fronts in large-scale weather systems, or by movement of air up the slope of a mountain or very large hill. Cumulus clouds are subdivided into *small cumulus, swelling cumulus,* and *cumulonimbus,* according to their degree of vertical growth.

Stratus Types

Stratus clouds, which are generally flat and horizontally layered, form over large regions as the result of gradual lifting for long periods in the vicinity of fronts associated with mid-latitude cyclones. The higher-altitude stratiform cloud types may spread many hundreds of miles from the cyclone center. Extensive stratus clouds can also be formed in air that is forced to rise over mountains or to gradually ascend more gently sloping terrain.

| Elevation and Precipitation | Stratus clouds are divided into categories depending on their height in the atmosphere—*cirrus* (*cirrostratus* and *cirrocumulus*) at high levels, *altostratus* and *altocumulus* at middle levels, and *stratus* and *stratocumulus* at low levels. |

Nimbus is a broad category of clouds from which precipitation is falling. Nimbus clouds are generally dark and amorphous in appearance and have ragged edges. A *nimbostratus* cloud often looks diffuse from beneath, but there may (or may not) be a cumulus-type cloud on its top side. *Cumulonimbus* clouds form heavy, dense columns as they rise from low altitudes to considerable heights.

Key to the Color Plates

The following illustrations appear with each text description opposite the color plates. They offer a quick-reference guide to the subject covered.

Symbol	Subjects	Symbol	Subjects
The Atmosphere	spacecraft and satellite views	Cumulus Clouds	cumulus, cumulonimbus, mammatus, gust front
Cirrus Clouds	cirrus, cirrocumulus, cirrostratus, contrails	Orographic Clouds 	mountain-induced cumuliform and stratiform
Middle Clouds	altocumulus, altostratus	Mixed Skies	cumulus and stratus combinations
Stratus Clouds	stratus, stratocumulus, nimbostratus	Rainstorms 	showers, thunderstorms, lightning, squall lines, microbursts

Symbol	Subjects	Symbol	Subjects
Tornadoes	mesocyclones, tornadoes, waterspouts, dust devils	Drought	drought, dust storms
Hurricanes	hurricanes, tropical storms	Vision Obstructions	fog, volcanic clouds, volcanic twilight
Snow and Ice	snowfall, blizzards, avalanches, ice storms, frost, freezing rain, sleet, hail	Optical Phenomena	auroras, halos, coronas, iridescence, rainbows
Floods	river floods, flash floods		

CLOUDS AND STORMS

Earth's Atmosphere

Two unique features of our planet from space are its blue color and the widespread and constantly changing clouds in its atmosphere. The blue color is caused by the scattering of the Sun's light by air molecules. Clouds are caused by a highly variable but significant amount of water vapor in the atmosphere. No two satellite views of Earth's clouds are ever exactly the same. This image features Africa in the center and Europe at the top.

Clouds The spiral cloud patterns of two large-scale traveling cyclonic storms can be seen over northern Europe (center top) and to the northwest (upper left) of Africa. Both systems have a small low-pressure center to the north and a cold front extending south and west. Along the equator are tropical waves that travel year-round from east to west. Over the cold South Atlantic (lower left) are unorganized clouds. A traveling low and a front can be seen off the southern tip of Africa. The counterclockwise circulation around lows in the Northern Hemisphere (reversed in the Southern Hemisphere) is caused by Earth's rotation.

The Atmosphere: Auroras

The thin skin that is our atmosphere is shown in this remarkable photograph, taken by an astronaut on the space shuttle *Discovery* in spring 1991 from a height of 160 miles (260 km). This low orbit placed the shuttle in the altitude range where auroras naturally occur, and in an orbit that took it close to polar regions.

The aurora is visible in two regions. One area is the bright vertical banding in the center of the image, where the aurora has a sharp lower boundary near the base of the ionosphere and grades upward along the magnetic field lines to an indistinct top. The other area is a fainter vertical draping in the lower left nearly underneath the shuttle. The aurora occurs at 60 to 80 miles (100–130 km) altitude in the ionosphere as high-energy particles from solar flares reach the earth and ionize gases so they emit a glow. A red aurora is due to emission by oxygen atoms. The ionosphere is a layer where many of the atoms that compose the atmosphere have lost electrons due to interactions with ultraviolet radiation, solar wind, and cosmic rays.

The Atmosphere: Hurricane Andrew

This satellite image shows Hurricane Andrew on August 24 1992, at 5:00 A.M., as it made landfall south of Miami. Andrew was the most costly natural disaster to strike North America in this century. It crossed Florida and the Gulf of Mexico, then became somewhat less intense when it hit the Louisiana coast. Total damages exceeded $25 billion; more than 100,000 homes were destroyed or damaged.

Clouds   This color-enhanced infrared image shows the hurricane's eye (its calm center) in red as it entered the Florida coast. The hurricane's cirrus cloud shield appears white over the eye wall (the tall bank of clouds that surrounds the eye and that contains the highest winds) and red on the outer edges. Lower clouds are green and blue.

Weather   Highest sustained winds in south Florida were 145 mph (233 kph), with gusts to 175 mph (282 kph). The storm surge reached 17' (5 m) on the southeast Florida coast. Most hurricanes cause as much destruction from water as from wind. Andrew had especially strong winds, which heightened the storm's enormous impact on coastal population centers.

HURRICANE ANDREW
24 AUGUST 1992
5 AM EDT   926 MB

The Atmosphere: Blizzard of 1993

This satellite image shows the blizzard of 1993 (or Storm of the Century) on March 13. As the strong and massive storm moved rapidly up the east coast of North America, it caused $1.6 billion in damage. If it had not been well forecast, the toll of people killed in the U.S., Canada, and Cuba—243—would have been much greater.

Clouds   Pale yellow areas in the Gulf of Mexico are tall thunderstorms penetrating the cirrus cloud shield, shown in black. To the north is an enormous yellow region of outflow from the storm and large-scale rising motion.

Weather   This was the first storm to close all major airports on the U.S. east coast. Heavy snow and blizzard conditions prevailed over a wide area. Snow reached 60″ (152 cm) in North Carolina and 20″ (50 cm) in Connecticut. There were tornadoes in Florida and a storm surge on the Gulf of Mexico. The surface low deepened explosively from the Gulf of Mexico to the Chesapeake Bay; all-time pressure records were set from South Carolina to Maine. Cold temperatures broke or tied records in 68 cities.

34

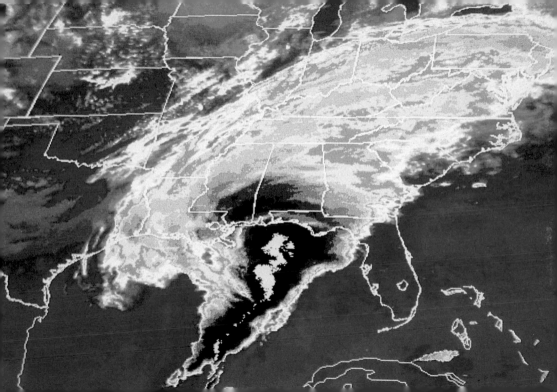

The Atmosphere: Cold Front

This satellite view, taken near sunset on November 15, 1989, shows a well-developed cold front extending from the Gulf of Mexico to the Great Lakes. Rather than taking the form of one continuous line, this front consisted of a series of separate storms. The infrared imagery depicts high (cold) clouds in red and yellow, and low (warmer) clouds in gray, white, and blue. Areas in red are high-level cirrus clouds—blow-offs of anvils that formed on top of cumulonimbus clouds, which grew in upward motions of air as the front progressed.

Weather   In late afternoon on this day, 39 tornadoes were reported. The strongest, in Huntsville, Alabama, killed 21 people and injured 463; property damage totaled $100 million. Behind the front to the west, lower clouds existed where cold air swept south. South of the front in warm moist air over the Gulf of Mexico, showers were disorganized but numerous.

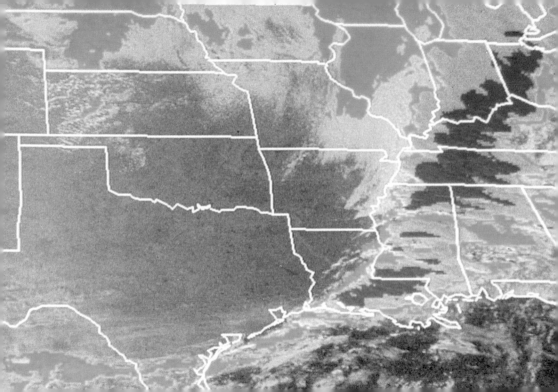

Cirrus

Detached, fibrous or wispy ice-crystal clouds, cirrus clouds typically occur above 16,000' (5,000 m) over most of the U.S. Most often, they are produced by upward motions in traveling large-scale weather disturbances. Sometimes they come from airflow over mountains or in regions of strong high-level winds.

Significance

A few thin cirrus clouds usually are not significant. An increasingly thick layer of cirrostratus may mean an impending increase in clouds or precipitation.

Season and Range

Cirrus clouds are common throughout North America at almost any time of the year. Satellite imagery sometimes shows cirrus blowing away from the tops of large-scale disturbances for thousands of miles, and lingering for hours after the storms dissipate. For this reason, not all cirrus clouds indicate nearby active weather. Ice crystals do not evaporate (or sublimate) nearly as quickly as water droplets, so ice-crystal clouds last longer than those composed of water droplets, such as altocumulus.

Thin Cirrus

Thin, white to light gray ice-crystal clouds, thin cirrus often occurs as streamers that cross but do not cover the sky. These clouds form when there is moisture at high levels and the temperature there is colder than freezing. Cirrus clouds are often elongated in the same direction across the sky. *Cirrus fibratus* clouds (mares' tails) have hooks on their ends. *Cirrus radiatus* clouds have long parallel streamers that appear to converge toward the horizon.

Significance

Since isolated cirrus clouds occur very frequently, they are usually not significant. Cirrus clouds accompany all large, organized storm systems, but they may be hidden by lower clouds. Increasingly dense and widespread cirrus clouds may indicate an approaching weather system.

Season and Range

All year. Wispy cirrus clouds are common across all of North America. In the drier southwestern U.S., they are seen quite often because the passing weather systems have fewer low clouds to block their view.

Cirrocumulus

Small but regular ripples of thin white clouds are unique to cirrocumulus, which often occur with other cloud types at higher levels. They represent moisture at high levels where the temperature is colder than freezing, and they occur in areas with above-average moisture or instability at the cirrus level. Cirrocumulus cloud features look small from the ground; if the ripples or waves are very large, the cloud is an *altocumulus* (middle cloud) or *stratocumulus* (low cloud).

Significance
Cirrocumulus clouds are most often found with larger-scale weather systems. A small or isolated cirrocumulus cloud is not especially important. Several patches or a widespread layer may indicate increasing moisture or instability ahead of an advancing weather system.

Season and Range
Never very frequent or widespread, these clouds occur year-round. They are usually mixed with other layered clouds at middle and upper levels across all of North America.

Cirrostratus

Uniform thin to thick, white to light gray ice-crystal clouds at higher altitudes, cirrostratus formations cover part or all of the sky. They help create a condition in which haloes around the Sun and the Moon may be visible. Cirrostratus clouds may not be spectacular, but they can indicate changing weather.

Significance
If cirrostratus clouds are becoming more dense and increasing, watch for an increase in the amount of other clouds over a period of several hours, since more active weather may be approaching. Cirrostratus clouds also may be part of an inactive high-level moisture outflow far from the source, such as from the southwest U.S. monsoon or from tops of subtropical or tropical disturbances.

Season and Range
Possible at any time, cirrostratus clouds are more widespread during winter in northern states with large-scale moving weather systems. In the southwestern U.S. they occur during the late summer monsoon, and in the southeastern U.S. during upper-level outward flows from subtropical and tropical disturbances.

Contrails

Straight, narrow lines of ice-crystal clouds typically produced by aircraft at altitudes colder than −25°F (−31°C), contrails occur in nearly saturated atmospheric conditions when aircraft emit water droplets that immediately freeze. The resulting ice crystals last longer than droplets. When they start to become diffuse, contrails can spread to form a short-term, localized cloud cover. Contrails often form when there are other cirrus clouds at the aircraft's level.

Significance   A few isolated contrails are quite common and of no significance. Their number is determined by the amount of air traffic overhead at higher altitudes. If the number of contrails and other stratiform clouds are both increasing, a large-scale weather system may be moving toward the region.

Season and Range   Contrails form when there is air traffic at higher altitudes with sufficient moisture.

Altocumulus

Thin or thick, white to gray, mainly water-droplet clouds at middle levels, altocumulus clouds often occur with variations that look like textured bands or patterned sheets. The appearance of altocumulus may vary significantly within a given cloud, and may change almost completely across the sky. Altocumulus clouds form with moisture at middle levels, and with temperatures slightly above to slightly below freezing. Altocumulus may occur in rows, organized patterns (*mackerel sky*), multiple layers, or small turrets, which indicate strong instability. The Sun may be visible through the cloud or may be hidden.

Altocumulus and stratocumulus are similar in shape, but if cloud features do not fill a large part of the sky, the cloud is likely to be altocumulus. Altocumulus clouds may occur at the same time as many other types of clouds. Altocumulus clouds exist above a height of 6,500′ (2,000 m); below that level similar clouds are considered to be stratocumulus.

Significance   Large areas of altocumulus usually accompany active, moving weather systems and the flow of significant moisture over hundreds of miles. When altocumulus clouds

last only a few hours or cover only a small portion of the sky, there is probably not an organized weather system in the area. If clouds are also at other levels, a more organized system is at, near, or approaching the area. When the amount and thickness of altocumulus is increasing overhead and changes are approaching from the wind direction at cloud level, moister weather is likely to be moving in aloft and perhaps at the ground also.

Season and Range   Altocumulus may occur anywhere, year-round, and for any of a wide variety of reasons. During winter in northern states and all Canadian provinces, altocumulus occurs near or ahead of moving weather systems. In the northern states during summer, altocumulus may come from similar systems or from large thunderstorms. To the south, altocumulus may come from northward flow from tropical and subtropical weather systems. Over western and Plains states during summer, moisture flowing from the southwestern monsoon can produce widespread altocumulus. In areas of the West with dry air and few low-level clouds, altocumulus is visible more often than in humid regions, where low clouds may block middle and high clouds.

Altostratus

Thin or thick, gray to pale blue, mainly water-droplet clouds at middle levels, altostratus are organized in gray to blue sheets without much texture or variation. The cloud sheet washes out the sky, giving rise to the term "watery sky." Altostratus clouds form with abundant moisture at middle levels, and with temperatures that range from a little warmer to somewhat colder than freezing. The Sun may be dimly visible through a thin, uniform altostratus layer. The height of featureless stratiform clouds is sometimes hard to determine.

Altostratus clouds lack the wisps or streamers of cirrus clouds, especially when an indistinct veil of precipitation falls from the altostratus base. If a middle-level cloud has variations across the sky, it is probably an altocumulus. If a stratiform cloud is rather smooth, it is more likely to be altostratus than stratus, since stratus clouds are lower and more likely to show variations.

Significance   Altostratus clouds are hard to identify and certainly not spectacular. If altostratus is increasing in coverage and approaching from the wind direction at the altostratus

level, then widespread precipitation may be expected in the region. This precipitation may not occur directly in the area, however, since altostratus clouds are often carried hundreds of miles without much change.

Season and Range

Altostratus may occur anywhere in the U.S. and Canada, year-round. During winter in the northern states and all Canadian provinces, these clouds occur far ahead of or near traveling large-scale weather disturbances. To the north during summer, altostratus clouds may come from similar systems. Altostratus may also be produced in summer by outflow at upper levels from large nighttime thunderstorms over the central U.S. To the south in summer and fall, altostratus comes from northward outflow from tropical and subtropical systems farther south. In the southwest and Plains states, altostratus is produced as moisture flows northward from the southwestern summer monsoon.

Stratus

Stratus clouds are generally diffuse and dull. They are low-altitude, light to dark gray water-droplet clouds that do not change much across the sky. This cloud formation has little structure and looks like fog, except that it is above the ground. Since stratus clouds may be quite close to the ground, they may have large and vaguely outlined features; if the cloud is very smooth, it is more likely to be the similar but higher altostratus. If the cloud has much variation across the sky, it is stratocumulus. Darker clouds and heavier precipitation come from nimbostratus.

From the vantage point of a large hill or mountain in the morning, stratus can be seen in a valley or lower area and appears similar to fog. The only precipitation from stratus is drizzle, light steady rain, ice crystals, or snow grains. The base of stratus ranges from only 100′ (30 m) up to 6,500′ (2000 m).

Significance   Stratus clouds indicate saturation near the ground. They can form locally overnight as a result of the cooling of Earth's surface, especially in lower valleys and depressions around lakes. They may also come from the flow of moist,

cold air into a region at low altitudes. Stratus may also represent the transition to or from fog and cumulus clouds.

Season and Range Stratus clouds may occur year-round, at any location with humid low-level conditions; they are common along coastlines and in regions with valleys. Along the Atlantic, Pacific, and Gulf coasts, low-level moisture is present or refreshed each day by onshore winds. Moisture also may be brought into a mountain valley and linger for days when winds are calm near the ground. Stratus often fills the low levels of an arctic air mass for several days anywhere east of the Rockies, especially in winter. Under all of these conditions, the air may reach saturation in the evening and form stratus clouds when winds are light. If the stratus clouds reach the ground, there is fog. Especially along coasts and in valleys, the fog or stratus often burns off partially or completely as the result of warming during the morning and early afternoon.

Stratocumulus

Stratocumulus are low, white to gray layered clouds with variations such as rows, lines, or patches. They are often arranged in bands or rolls that lie across the wind. Light rain, snow, or sleet may fall from stratocumulus. These clouds represent saturation and instability in a shallow layer of air near ground level. The bases of these clouds are a few hundred feet to 6,500' (2,000 m) high.

Significance
Stratocumulus may be a recurring local condition along a coast at certain times of the year. In other locations, stratocumulus clouds accompany larger-scale, traveling weather systems, typically after the passage of a cold front or other disturbance, and especially in the afternoons. Stratocumulus clouds are often seen before the sky clears, or they may be the final result of the decrease in cloud cover before a new weather cycle begins.

Season and Range
In the northern U.S. and Canada, stratocumulus clouds form year-round after the passage of humid large-scale disturbances; to the south, the same occurs in wintertime. South, southeast, and east of the Great Lakes, cold air passing over the warmer water produces stratocumulus.

Nimbostratus

Dark gray to pale blue and formed from water droplets, nimbostratus are rain and snow clouds. Their blurred, ill-defined look comes from falling precipitation. In winter, they may give the sky a smooth and indistinct appearance. In summer, these rainclouds often cover only part of the sky; blue sky and the full shape of other clouds can be seen as well. Nimbostratus is a deep cloud that appears to have dim and uniform lighting emanating from within. It may be similar to thick altostratus, stratus, or stratocumulus, but is distinguished by producing rainfall. When lightning or thunder occurs along with rain from a nimbostratus cloud, the cloud has become a *thunderstorm*.

Significance   Nimbostratus clouds are important to watch, as they always bring precipitation. If a nimbostratus cloud is convective (growing vertically through the upward motion of warm air), a thunderstorm may develop rather quickly.

Season and Range   Nimbostratus clouds are visible anywhere, year-round. In the eastern and southern U.S., fog, haze, and other suspended particles may make it hard to tell if a cloud has precipitation until it is nearly overhead.

Small Cumulus

Cumulus are fair-weather, water-droplet clouds that are detached from one another. They generally have well-defined, flat bases and domed tops resembling cauliflower. Their outlines are sharp, and they often develop vertically in the form of rising mounds, domes, or towers. Their sunlit parts are brilliant white; the base is relatively dark and roughly horizontal. Under normal conditions, the bases of all the cumulus clouds in one area of the sky form at the same altitude, which is the condensation level of the rising air currents, so that, when viewed at a distance, their flat bottoms seem to merge into a level plane.

Small Cumulus Clouds

No taller than they are wide, small cumulus clouds have bases that are generally flat with some darker areas, while their tops are somewhat rounded. Rain does not fall from small cumulus clouds; they are too shallow to produce droplets in the short distance from cloud top to base. Small cumulus clouds are caused by rising currents of warm air that cool to the dew point and condense. They are very common in the U.S. and Canada during the warm season, and at any time of year when there is an adequate supply

of low-level moisture and vertical motion from large-scale disturbances, coastlines, or mountains and hills. When cumulus clouds exist in areas with high wind speeds, they can be torn into nearly horizontal fragments with shallow depth. Small cumulus sometimes grow into larger cumulus clouds; when a cloud's size approaches a depth that is greater than its width, it is in transition to the next stage, swelling cumulus (cumulus congestus). The bases of cumulus range from about 2,000′ to 4,000′ (600–1,200 m) in humid regions to as high as 8,000′ to 15,000′ (2,400–4,500 m) in dry regions.

Significance   Small cumulus clouds are a sign of generally good weather if they do not grow much. However, the formation of shallow small cumulus clouds within a blue sky may be only a temporary stage, as some continue to grow into larger cumulus. When continuous upward currents of warm air are present, small cumulus can become swelling cumulus. When small cumulus occur in the morning throughout midday, the atmosphere may be unstable enough for them to grow into larger clouds later in the day. If small cumulus are the only convective clouds (those caused by rising small-scale air currents) during the afternoon, there is no

threat of showers in the short term. In some areas and situations, small cumulus may begin to grow at or after sunset and develop into large nighttime thunderstorms. Even though you may observe an abundance of small cumulus clouds, there also may be thunderstorms and cumulonimbus in other portions of the sky.

Season and Range

Small cumulus with low cloud bases are very common, especially in summer in the eastern U.S. and Canada and at any time of year along the Pacific coast. Large-scale traveling disturbances in summer can produce vast regions of small fair-weather cumulus clouds away from areas of strong upward motion.

In the southeastern U.S. during warm months, small cumulus start the cloudiness cycle nearly every day due to the presence of both adequate low-level moisture and temperatures that are warm enough to form updrafts that produce cumulus; small cumulus may be only a temporary stage as some clouds continue to grow into cumulus congestus or cumulonimbus. Higher bases are typical during the afternoon in summer in the western states, and during the afternoon in summer in the northern half of the U.S. and Canada during less humid periods.

Swelling Cumulus

Cumulus clouds swell into tall towers when upward currents of warm air continue beyond the small cumulus stage. Swelling cumulus—also called *towering cumulus* and *cumulus congestus*—are taller than they are wide. Tops of swelling cumulus clouds have hard outlines; if they are fuzzy or fibrous, ice is starting to form (that is, the clouds are moving to the cumulonimbus stage).

Sometimes swelling cumulus clouds tilt because the wind has different speeds at the top than at the cloud base. They often are separate clouds or may be somewhat organized. The largest of these white water-droplet clouds have some darker areas at their bases or on their sides, and they often have light rain falling from them. For a cumulus to reach the congestus stage, the atmosphere must be somewhat unstable in a deeper layer than is required for small cumulus. And the congestus stage may not be the last stage of growth; the cloud may continue growing and become a cumulonimbus. If there is lightning or thunder, the swelling cumulus is considered to be a cumulonimbus. The bases of swelling cumulus range from about 2,000′ to

70

4,000' (600–1,200 m) in humid regions to as high as 8,000' to 15,000' (2,400–4,500 m) in dry regions.

Significance   Swelling cumulus should be watched over a period of time to see whether they are the largest cumulus that will occur all day, or if they will develop into cumulonimbus. The earlier in the day that congestus clouds form, the more significant they are as indicators of strong storms later that day. If there are quite a few swelling cumulus clouds on a sunny summer day in the morning to early afternoon, the atmosphere may be unstable enough for them to reach the cumulonimbus stage, which can produce significant rain or severe weather. But if cumulus congestus are the largest cumulus clouds that have appeared by late afternoon, they probably will not grow any larger for the rest of that day. However, other subtle factors may help them grow further just before or after sunset; be aware of any evening strengthening of swelling cumulus before deciding that the risk of cumulonimbus is over for the day. Note: cumulus congestus may cover one part of the sky while there are small cumulus in one direction and huge cumulonimbus in another direction. In tropical weather, such as Florida in summer, cumulus congestus may

produce heavy rain showers for a few minutes, or form into cloud lines that last for several hours.

Season and Range

Mainly summer. In the eastern U.S., cumulus congestus with low cloud bases are fairly common during warmer months. In the drier western U.S. and Canada, afternoon heating causes updrafts that form clouds with higher bases over large elevation changes around mountains. Along the west coast, swelling cumulus clouds occur occasionally during warmer months. They also may be associated with upward motions from large-scale traveling weather systems, such as to the south and west of cold fronts. Congestus clouds often form on summer afternoons within the small zones of upward air motion that are produced daily by breezes near oceans or the Great Lakes, and by mountain breezes that occur where there are large vertical changes in land surface. Lake and sea breezes may move the clouds tens of miles inland, or occasionally a few miles offshore, depending on upper-level winds. During colder months downwind from the Great Lakes, swelling cumulus clouds sometimes form in long lines and produce heavy snow; the tops of these clouds may not exceed 10,000′ to 15,000′ (3,000–4,500 m).

Cumulonimbus

Cumulonimbus are isolated to highly organized clouds made up of water droplets in their lower portions and ice particles in their upper portions; they have dark bases. Precipitation always falls from cumulonimbus clouds, and severe weather most often occurs with them; if there is lightning or thunder, the cloud is considered to be a cumulonimbus. In arid regions, the rain or snow may evaporate before it reaches the ground; this phenomenon, called *virga,* is often accompanied by gusty winds.

A cumulonimbus is formed when rising currents of warm air cool to saturation; water vapor in the currents condenses into droplets, forming a small cumulus. Continued rising warm air currents create a swelling cumulus and, finally, a cumulonimbus. For this stage to be reached, the atmosphere must be very unstable in a deep layer. Cumulonimbus clouds are deep enough that cloud bases often are not visible at a distance when the rest of the cloud can be seen. Bases of cumulonimbus range from about 2,000' to 4,000' (600–1,200 m) in humid regions to as high as 8,000' to 15,000' (2,400–4,500 m) in dry regions.

The Anvil

Two important visible features of a cumulonimbus cloud are the icy top and the anvil. When a cumulus has a fuzzy and fibrous top instead of a hard outline—indicating that ice is present—a cumulonimbus has formed. A cloud also qualifies as a cumulonimbus when part of the top is flat like an anvil or has an overhang above clear sky. Sometimes the anvil blows a long way downwind at cirrus levels; detached anvils often give rise to cirrus clouds in subtropical and tropical regions and during warmer months elsewhere.

A cloud with an anvil is always a cumulonimbus. The anvil shows that the cumulus has stopped growing at a high level, usually above 30,000′ (9,000 m), because of strong stability—warming or drying—above the anvil's top. Anvils may blow off for a long distance and become detached or much larger than the parent cumulus cloud. Sometimes the anvil may partially detach or be the only part of the cumulonimbus that remains. A full cumulonimbus and anvil structure may also be formed in a cold winter environment. Some cirrus clouds formed by anvils are thick enough to be gray on the side away from the Sun.

Pileus

A pileus is a small horizontal accessory cloud—at a high level and quickly changing in form—that often occurs

above a cumulus or cumulonimbus; it may appear as a hood that is draped across the top of the cloud tower beneath it. A pileus is formed when a relatively shallow layer of moist air is lifted by a strong updraft and cooled below its dew point. Most updrafts that are strong enough to produce a pileus cloud will also result in a cumulonimbus. A pileus may become draped over a cumulus cloud as the latter grows to the cumulus congestus or cumulonimbus stage.

Significance   Called *Cb* in the aviation and meteorological communities and abbreviated as such on weather maps, cumulonimbus are very important clouds to monitor whenever there is a good supply of low-level moisture and strong upward forcing of the air. The earlier in the day that a cumulonimbus forms, or the more organized and vigorous that it appears, the more likely it is to become significant in terms of storms and weather.

When cumulonimbus becomes arranged in rows or complexes, the likelihood of severe weather increases. Severe weather that can accompany cumulonimbus includes tornadoes, waterspouts, and funnel clouds; brief to prolonged heavy rain, hail, sleet, snow, and flash floods; strong winds and turbulence at any level of the

atmosphere, including gust fronts, squall lines, and microbursts; and lightning from cloud to ground, within clouds, and between clouds.

Season
and Range

These clouds are much more common during warm months. In the eastern U.S., cumulonimbus with low bases form most often in spring and summer. In humid areas, adequate moisture and instability aloft—the requirements for the formation of a cumulonimbus—often occur near cold fronts and large-scale traveling systems. Over the larger mountain ranges of the dry regions in the western U.S. and Canada, summer cumulonimbus clouds occur with strong afternoon heating over or along mountain slopes, more vigorously when there is moisture aloft.

Along the west coast, dry and sinking air aloft inhibits most cumulonimbus growth, even though surface air may be humid. Cumulonimbus with an anvil may also form during cold weather in areas near water, for example, in the lee of the Great Lakes (that is, downwind from the lake); such clouds often occur in lines and produce heavy snow.

Mammatus

A series of pouch-like gray to pale blue clouds hanging down from middle or upper clouds, mammatus vary in size and are found under the anvil blow-off from a cumulonimbus. (The cumulonimbus may be too far away to be visible.) They are often seen in the vicinity of severe thunderstorms. The individual elements of mammatus represent inverse convection—that is, instability in the downward direction—caused by the large amount of moisture and heat flowing out from a cumulonimbus anvil over the relatively undisturbed and drier air below.

Significance
When mammatus are weak, in the distance, or not approaching, the parent storm is not likely to be very important. Under other conditions, the sky should be watched for possible severe weather. Mammatus clouds most often follow the most active growth stage of a cumulonimbus within minutes to an hour or more.

Season and Range
Since they result from cumulonimbus, mammatus are most common during warm months: spring and summer in the eastern U.S., and summer over the mountains of the western U.S. and Canada.

84

Gust Front

All cumulonimbus clouds form outflows that travel away from the parent thunderstorm's rainfall. Gust fronts are arc-shaped low clouds on the leading edge of cooler air flowing out from a cumulonimbus. A gust front may be a few hundred feet off the ground in humid regions to 15,000' (4,500 m) high in dry areas. It may have an anvil of cirrus clouds overlying it at higher levels and flowing in the opposite direction. Gust fronts occur more often in partial, semicircular shapes than in long, organized lines, but they can become highly organized into a continuous, curved line, up to 100 miles (160 km) long, marked by low cumulus or stratus clouds.

Significance   Wind typically increases in speed and changes direction when a gust front passes, creating an aviation hazard. Winds greater than 40 mph (65 kph) over large areas are typically from *squall lines,* which consist of strong gust fronts organized together. Winds may increase up to 15 minutes after the passage of a gust front.

Season and Range   Gust fronts occur during warm months in the eastern U.S., less often during summer in dry and mountainous areas.

Mountain-induced Cumulus

When the atmosphere is unstable vertically, moist air flowing over mountains or very large hills sometimes is forced upward and forms clouds with cumulus shapes. In such a process, called *orographic lifting,* the cloud is relatively stationary over the mountain or hill that caused it to form. Depending on the height of the moist layer, *orographic cumulus* may form as wind flows up the mountain, near the top (sometimes in the form of a banner cloud extending downwind), higher than the mountaintop, or several miles downwind from the mountain. These clouds may appear stationary, but individual cumulus are forming, growing, and dissipating over a period of tens of minutes. As with other cumulus, these are water-droplet clouds with rounded tops. Occasionally a wall of clouds with some imbedded cumulus shapes may develop over a mountain ridge. Cumulus and stratiform orographic clouds can exist simultaneously.

Mountains induce complicated, changing wind-flow patterns throughout the day, so orographic cumulus clouds on one side of a mountain in the morning are not likely to

stay there all day. An exception is in Hawaii, where steady trade winds form orographic cumulus clouds that remain for hours on one side of a mountain.

Significance   Orographic cumulus contain areas of strong winds and the chance of increasing wind speed inside the cloud and sometimes on the ground beneath the clouds; as such they require caution on the part of operators of aircraft. In order for these cumulus clouds to form, there must be enough moisture in a layer of air for cumulus to develop by normal processes. Sometimes thunderstorms may grow later in the day if wind flow, stability, and moisture conditions are suitable for the further growth of the clouds to the cumulonimbus stage.

Season and Range   All year, but more common in summer. Orographic cumulus form over mountains of the western states and provinces in the presence of adequate moisture aloft. They also occur over the mountains and largest hills (at least 500'/150 m high) of the eastern U.S. and Canada.

Mountain-induced Stratiform

Stationary, smooth-edged clouds, called *orographic stratus clouds,* form as wind flows up a mountain or large hill and down the other side. There must be enough moisture at or above the highest ground to condense into water droplets. Stratiform clouds are shaped as waves, lenses (lenticular clouds), stacked plates, rolls, veils, layers, or a combination. They may sit in a cap or crest at the top, or form at higher than mountaintop level, or much higher yet in the lenticular shape. Form and position depend on the height of the moisture layer and the vertical variations in the speed and direction of the wind flowing through the cloud. Stronger upward motions occur when winds blow perpendicular to the mountain ridge line rather than at an angle. While the clouds appear stationary for hours, portions form, grow, and dissipate every few minutes. Skies around mountain-induced stratiform are very clear because the airflow is usually sinking from very high altitudes where there is no moisture or pollutants. These clouds are destroyed when passing weather systems change the wind direction and speed, temperature, and moisture near cloud level.

When the wind flow is perpendicular to the barrier it is blowing against, the air often warms as it flows downhill, such as in the westerly Rocky Mountain chinook and the northerly Santa Ana winds of California. Such winds occur mainly over and close to the mountain ranges; the wind may reach gust speeds of 150 mph (240 kph) and maintain its strength for several hours, most often at night. When it blows from the north-northwest, as in the Rocky Mountain bora, the air stays cold; bora winds may spread onto downwind plains up to 50 miles (80 km) away.

Significance   Orographic stratiform clouds indicate strong winds and turbulence inside the clouds and sometimes on the ground directly beneath them. Winds blowing as high as 100 mph (160 kph) or more in stratiform orographic clouds are a potential hazard for aircraft.

Season and Range   Primarily in winter months. Orographic stratiform clouds develop regularly over mountains of the western states and provinces and occasionally over mountains and large hills of the eastern U.S. and Canada. The terrain must have a fairly abrupt elevation change of at least 500' (150 m) in humid regions, higher in more arid areas.

Mixed Skies

The sky is in a constant state of flux, exhibiting many stages of cloud growth and development throughout the day. The cloud types described in this book are the components of mixed skies—the common scenario in which more than one cloud type occurs overhead at different levels of the atmosphere. Wind speed and direction, temperature, and moisture content vary from layer to layer and among clouds of various sizes and types. The mixture of clouds in the sky—and the ways they grow, transform, blend together, and dissipate— indicate the causes of the present weather.

Beyond an understanding of current conditions, the combination of clouds usually provides clues about what is likely to occur as the day progresses. No two skies are ever exactly alike. To observe clouds and decipher their messages regarding impending weather, you must watch the sky periodically and frequently over the course of the day.

Mixed Cumulus   At any one time, cumulus clouds typically exist in several stages of development. Since all cumulus start to form

at cloud base and grow upward, there are likely to be small cumulus if larger ones are around. On a warm summer day there are likely to be small cumulus, swelling cumulus, and cumulonimbus both far and near.

When planning an outdoor activity, the most important feature to watch for is the anvil from a cumulonimbus. The anvil, typically shaped as an overhanging cloud segment protruding from the cloud's fuzzy or fibrous icy top, forms when the cumulonimbus is deep and therefore capable of significant or severe weather. The stage leading up to the cumulonimbus is the swelling cumulus, a cloud with hard tops on tall towers. Sometimes it is hard to see the critical clue that a cumulonimbus is present, or that a swelling cumulus is exploding upward from its formerly harmless small cumulus stage. Watch the field of cumulus clouds over a period of time and look for a very tall tower with a hard top or for a small anvil, or when the cumulus cover the sky, look for a darkening and more solid appearance to the cloud bases. Scan the entire sky, especially near the horizon, for new developments.

Mixed Stratus   When there is a stratus cloud at one level, more may be present elsewhere in the sky. In this book stratus clouds

are divided according to their altitude—high (cirrostratus and cirrocumulus), medium (altostratus and altocumulus), and low (stratus and stratocumulus). Stratus clouds can coexist at several elevations because they frequently are the result of gradual upward motions in large-scale traveling disturbances. These weather systems typically have rising motion through much of the depth of the troposphere. Throughout the day, or night, layers with enough moisture to be close to saturation may reach their dew point as the air is lifted.

Stratus clouds can also occur singly. A cirrus layer cloud may occur without other stratus clouds; often one blows in from a distant location. Other single-layer types that commonly form alone are stratus or stratocumulus from nighttime cooling at low levels on a coast or in a valley. When stratus clouds do not extend across the entire sky, it can be easy to tell which layer is above or below another layer. But when low stratus covers the sky, middle or high clouds at higher altitudes may not be visible. During takeoff from an airport it is not unusual to climb above the lowest layer only to find one or more layers above it, especially if precipitation is falling.

Mixed Cumulus and Stratus

The significance of mixed cumulus and stratus depends on the development stage of the cumulus, and the amount and coverage of layer clouds. Small cumulus and thin cirrus are generally not important. But as larger cumulus occur with more layer clouds, more significant weather is possible.

Layer clouds at any level reduce the heat reaching the ground. Cooling depends on cloud thickness, the amount of sky cover, the number of layers, and how long clouds persist. In turn, cooling of the ground by layer clouds affects how far the cumulus will grow upward. When cumulus reach the swelling or cumulonimbus stage, there may well be significant weather, despite substantial cloud layers.

In subtropical and tropical regions, such as Florida or Arizona during the monsoon, multiple layer clouds and cumulonimbus usually accompany the very heaviest rains. Cumulonimbus arranged in long lines and embedded in thick layer clouds indicate a strong likelihood of heavy rains, high winds, and other severe weather. The most extreme case of this phenomenon is the hurricane rainband.

102

Showers and Thunderstorms

A shower has no thunder or lightning and usually lasts less than 15 minutes. Brief, light showers may come from stratocumulus, but showers and thunderstorms typically originate from a cumulonimbus formation. In the dry western U.S. and Canada, *virga* (when precipitation from showers or thunderstorms evaporates before reaching the ground) sometimes occurs. High winds are more likely when virga occurs than if heavy rain reaches the ground.

Thunderstorms can produce a great variety of weather phenomena—rain, hail, strong, gusty winds, sudden temperature changes at the ground, lightning, thunder, and even tornadoes. Within a few minutes, a thunderstorm can make a still and oppressively close summer day windy and comfortably cool. Thunderstorms are actually quite rare. At most locations in North America, they are in progress during less than 1 percent of the hours in a year.

Formation   The essential ingredient in the creation of a thunderstorm is a warm, moist, unstable atmosphere. This necessary instability doesn't happen all that often. Normally the atmosphere is *stable;* a good example is the cold air at the

bottom of a valley. Cold air is "heavier" or denser (meaning it has more air molecules per cubic foot) than warm air. When lifted off the ground by, say, 100 feet, the cold valley air will find itself surrounded by warmer, "lighter" air and will immediately sink back to the valley floor. Rising currents do not go very far in stable air, and incipient cumulus clouds have little chance to grow into thunderstorms. In *unstable* air (air that tends to continue to move upward), temperatures decrease rapidly with height—about 5.5°F per 1,000 feet. Although a rising bubble of air expands and cools as it goes up, it remains warmer than the surrounding air and keeps on rising. Instability is even stronger if there is water vapor in the air. The vapor releases its latent heat as it cools and condenses into droplets, and the release of the heat further supports the rising motion.

Life Cycle   Showers and thunderstorms come from cumulus convection, so they grow, mature, and dissipate rather quickly. During a thunderstorm, the flows of wind in both horizontal and vertical directions are much stronger and more turbulent than in large-scale traveling weather systems. In the *growth stage,* as a thunderstorm explodes,

strong upward motions may induce tornadoes, funnel clouds, mesocyclones, wall clouds, and waterspouts. During the early portion of the *mature stage,* there may be strong downdrafts in the form of gust fronts, outflows, and microbursts. During the *dissipation stage,* the cumulative effects of continuing rainfall may cause flash flooding. Often this stage occurs during the late night or early morning hours. Gust fronts may move outward from a dissipating thunderstorm complex and travel long distances before dissipating. Lightning can occur during any of the stages of a thunderstorm.

Effects   Although showers and thunderstorms can cause death and destruction, their effects are also beneficial. Much of the rainfall during the growing season in the central U.S. comes from large groups and complexes of showers and thundershowers. As many occur overnight, their benefits are enhanced by reduced evaporation. Lightning oxidizes atmospheric nitrogen, putting that essential plant nutrient into a form that is much more readily absorbed into the soil.

Lightning

When liquid water droplets within a cloud freeze into ice, electrical charges develop. Ice crystals become positively charged, while the remaining liquid droplets take on a negative charge. Soon the whole cloud becomes charged: positive charges collect in the icy cloud top; negative charges accumulate in the lower, warmer parts of the cloud. Normally the ground is also negatively charged, but the concentration of electrons in the lower cloud repels the negative ground charge (like charges repel; opposites attract), leaving positively charged ground directly beneath the cloud. The gathering electrical charges build voltages as high as 100 million volts within the cloud and between cloud and ground. Air can separate voltages as great as 3,000 volts per foot, or 15 million volts per mile, but when voltages exceed these values, lightning results.

Sequence   Lightning begins as a relatively weak and faintly visible *leader stroke* makes its way down from a cloud base to (1/100th second later) a tree, the ground, or several targets, completing an electrical pathway between cloud and ground. A massive *return stroke* shoots up along the

leader path at one-sixth the speed of light. Return strokes from several ground targets may join several hundred feet up, forming *branched lightning.* The concentration of electricity in a path 1" (2.5 cm) across heats the air almost instantaneously to tens of thousands of degrees (hotter and brighter than the surface of the Sun). We see the glowing channel as a cloud-to-ground lightning flash, while the sudden heating and expansion of the air makes *thunder.* Other varieties, such as *in-cloud lightning* and *cloud-to-cloud lightning,* discharge in a similar manner. Many more cloud lightning discharges occur than flashes to the ground; most often—especially at night—lightning is seen as illumination inside a storm.

Significance | Distant lightning indicates instability sufficient for a thunderstorm to develop. Lightning that is close enough for thunder to be heard indicates that a storm may be imminent. Sound travels at 5 seconds per mile (3 seconds per km). When a lightning flash is seen and its thunder follows 15 seconds later, the lightning is 3 miles (5 km) away.

Season and Range | Lightning occurs year-round in warm climates, and mainly in summer where winters are cold.

Squall Lines

A squall line is a continuous line of thunderstorms or showers, ranging from 20 miles (32 km) to more than 1,000 miles (1,600 km) in length; it is usually moving fairly rapidly. A squall line forms when outflows from a cold front or group of individual cumulonimbus clouds organize into a long line; it is often ahead of the cold front by tens to hundreds of miles. The lower leading edge of a squall line may appear in the shape of a smooth shelf cloud. As a squall line passes directly overhead, the dark and turbulent cloud base becomes visible. At that time or shortly after, strong winds and heavy precipitation begin; these are often brief. Later the cold front will pass, with winds shifting to come from a more northerly direction, and temperatures will fall. A low shelf cloud that is semicircular and located on the edge of rain indicates the presence of a gust front from a single cumulonimbus rather than an entire squall line. A squall line usually moves at a minimum of 30 mph (50 kph); winds can exceed 100 mph (160 kph) over long sections of the line.

Significance   For a squall line to be maintained, moist air needs to flow into it continually along its length. Squall lines are usually on the side of a front or thunderstorm complex in which moist, warm air flows toward the line and rises. Thunderstorms with squall lines are likely to be intense, often with damaging straight-line winds, hail, and hard but brief rainfall. The straight-line winds blow from the same direction along the entire line, usually west to northwest. Small, short-lived tornadoes may occur along the leading edge of a squall line.

Season and Range   In the U.S. and Canada, squall lines are on the south to east of fronts, where inflow air meets cooler outflow from north to west. The strongest squall lines occur in spring and early summer over the Great Plains from Texas to Minnesota, where strong cold fronts exist, while air to the south and east is moist and warm. In the southeastern states, squall lines may occur during any month of the year; they are often not as organized as the Plains systems. Most squall lines in the Pacific states and provinces occur in winter. They are rare over the mountainous region of western North America due to the lack of low-level tropical air.

Microbursts

A microburst is a sudden, short-lived, localized wind that often appears to radiate outward from a central point. Microbursts rarely occur in their classic form, but portions or variations of them are frequently seen. Classic microbursts are visible as a descending plume of rain or dust that spreads horizontally when it reaches the ground. In extreme cases, dust or rain may curl from the ground back up toward the cloud as downdrafts blast the ground at speeds up to 150 mph (240 kph). Microbursts can spread outward in a circle, but winds usually produce fan-shaped patterns in one direction. A microburst lasts 1 to 5 minutes and covers an area less than 2.5 miles (4 km) in diameter. A *downburst* is a strong downdraft-induced wind over a larger area. Both types come from a single swelling cumulus or cumulonimbus. Microburst winds descend and spread outward, unlike tornado winds, which converge and rise.

Effects Microbursts were first identified because of the major aircraft disasters they caused. Special types of radar are being installed at many airports for identifying these dangerous phenomena. Training programs for commercial

pilots describe the impact of microbursts on aircraft performance and explain avoidance procedures, but private aviation continues to be highly vulnerable. Microbursts are also a hazard to boating, and they can fan forest fires in unexpected directions.

Dry Microbursts

Dry microbursts, which occur over the semiarid western Great Plains and the mountain regions of western North America, typically emerge from a swelling cumulus or cumulonimbus cloud with a high base (10,000'/3,000 m or more) that has developed in moist air at middle levels. Most of the precipitation from dry microbursts evaporates before reaching the ground, and the evaporative cooling intensifies the downdraft in the dry low-level air.

Wet Microbursts

Wet microbursts occur in extremely wet environments east of the Rocky Mountains. These wet downdrafts have nearly saturated lower levels, but there is drier air at middle levels outside the storm; the dry air fuels the evaporative cooling that is necessary to accelerate the downdraft. Wet microbursts are often embedded in heavy rain; they are sometimes associated with tornadoes and larger-scale squall lines and gust fronts.

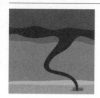

Mesocyclones

A mesocyclone, or *supercell thunderstorm,* occurs when an entire cumulonimbus cloud column, or a portion of one, rotates from the cloud base upward. Typically a few miles across, this rare type of storm may spawn strong tornadoes or other extreme weather. To describe it another way, a cumulonimbus formation may consist of a rotating cloud column (or may contain a rotating portion) called a mesocyclone, and a mesocyclone may contain a tornado. Mesocyclones require a very special set of meteorological conditions. Principal are a very moist low-level air mass and drier air aloft, extreme vertical instability and strong updrafts, a strong jet stream aloft, and sharp changes in wind speed and direction from south to southwest.

A classic mesocyclone has a main cumulonimbus with a flat *rain-free base* where strong updrafts are found. The *wall cloud* is a distinct and often rotating lowering of the rain-free base from which a tornado may drop within minutes. A *flanking line* of swelling cumulus clouds may extend southwest from the cumulonimbus. When a mesocyclone

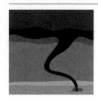

occurs, it is typically at least 80°F (26°C) at the ground and relatively cold aloft. At the ground, strong southerly winds bring in very warm moist air. A few thousand feet (about a kilometer) higher, winds are more from the southwest. Above 20,000' (6 km), winds are very strong from the west in the jet stream.

Significance A mesocyclone will often produce severe weather, such as strong winds, large hail, and tornadoes; it is the most intense portion of a thunderstorm. In the dry western Plains states, the entire cumulonimbus may be a mesocyclone. Such a storm is tall, isolated, rotating, and has smooth, rounded walls; it has light precipitation but can produce tornadoes. In regions with heavy precipitation, a mesocyclone can be hidden by rain and difficult to identify—this is why radar identification is so important.

Radar Detection The network of Doppler radar being installed across the U.S. during the 1990s can detect the cloud rotation that identifies a mesocyclone. Once rotation is detected, severe weather warnings are issued by the National Weather Service. Strong mesocyclones may rotate above the cloud base within the cumulonimbus for up to 20 minutes before circulation deepens and a tornado touches ground.

Tornadoes

The most violent storm that can occur at a given point, a tornado is a rapidly rotating column of air extending from the base of a swelling cumulus or cumulonimbus cloud down to the ground. A *funnel cloud* is an incipient tornado that has not reached the ground. Most tornadoes have a funnel cloud phase before and after the tornado stage. Some funnels never become tornadoes. A tornado's narrow, rotating spiral of air is usually larger at cloud base than at the ground. The column is made visible by the condensation of water vapor, or by the presence of dust or debris raised from the ground. Winds on the surface of the column are usually spiraling upward and can reach 250 mph (400 kph) in very small areas. Almost all Northern Hemisphere tornadoes spin counterclockwise.

The powerful updrafts inside a tornado can suspend tremendous loads of dirt and debris. When the tornado weakens or dissipates, the debris cloud may suddenly collapse toward the ground, spreading horizontally away from the dissipating tornado, much like a microburst.

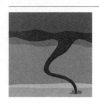

Shape   The classic shape of a tornado is a cone that narrows from cloud base toward the ground, but it may be obscured by dust and whirling debris at the surface. Very large, destructive tornadoes may appear only as a thick amorphous mass of menacing black clouds near and on the ground. Very small tornadoes can be thin, writhing, ropelike columns of gray to white clouds and dust. In drier air and during weaker tornadoes, the condensation cloud may not be complete between cloud base and ground; there may be a funnel coming down from the cloud and a dust whirl rising from the ground directly beneath it—any part of the cloud between the upper funnel and the debris cloud may be transparent or invisible. However, the effects of tornadic winds can be the same even if the cloud is incomplete.

Narrow streamers of rain at a distance can look like tornadoes, but may not have all of the following features: sharp vertical edges, horizontal travel along the ground, rotation, and an elongated, tapering shape. Turbulent and threatening *scud clouds* (cloud fragments beneath the main cloud layer) and low-hanging elements on horizontal shelf clouds from outflows are sometimes erroneously reported as tornadoes, but both lack rotation.

Formation

The largest and most dangerous tornadoes develop from parent mesocyclones (supercell thunderstorms). Both require the same conditions: moist, warm low-level air, dry air aloft, strong instability, a jet stream, and southerly winds at lower levels, westerly winds at upper levels. Powerful updrafts inside the supercell thunderstorm draw in the slowly rotating air circulating around it, concentrating the spinning motion. As the updraft strengthens, the spinning increases, until the updraft becomes a narrow, rotating column. Large outbreaks of violent tornadoes usually occur east of upper-level traveling disturbances. Less intense tornadoes can occur near squall lines and gust fronts, beneath rapidly swelling cumulus and cumulonimbus, and within hurricanes.

Non-supercell tornadoes, or *land spouts,* form when updrafts in a rapidly developing cumulus congestus or cumulonimbus cloud draw in slowly rotating low-level air. The preexisting slow rotation of the low-level air can be caused by airflow around mountains or ridges or the convergence of sea breezes or gust fronts. The rotation of many non-supercell tornadoes begins near the ground and grows upward.

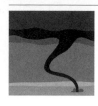

Movement and Intensity

Most tornadoes move from southwest to northeast. But they may come from other directions, form a loop, or be stationary. Typical speeds are 35 mph (55 kph), but some move at up to 70 mph (115 kph). Any tornado can cause significant damage. A typical tornado lifetime is less than 15 minutes, but some have lasted up to 7 hours.

Tornado intensities are rated with the Fujita-Pearson Scale. More than half of all tornadoes are rated as *weak;* their winds are typically below 110 mph (180 kph), their path lengths are shorter than 3 miles (5 km), and their path widths are narrower than 55 yards (50 m). Only 2 percent of tornadoes are rated as *violent,* but the winds of these storms can reach 250 mph (400 kph); their path lengths can exceed 300 miles (500 km), and their widths 3 miles (5 km). *Strong* tornadoes attain values between these two categories.

Over the past century, tornadoes in North America have killed between 15,000 and 20,000 people, and injured many more. During the 1980s, human loss in the U.S. and Canada averaged about 60 per year. Property loss from tornadoes averages at least $1 billion a year.

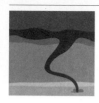

Season and Range

The Great Plains of North America provide the most favorable environment on Earth for the formation of tornadoes. Seventy-five percent of the world's reported tornadoes (about 1,000 annually) occur in the U.S., and another 5 percent touch down in Canada. Most tornadoes occur east of the Rocky Mountains in spring. States along the Gulf of Mexico start the tornado season in early spring. During April and May, the highest numbers of tornadoes occur in Texas, Oklahoma, Kansas, Nebraska, and the Dakotas. In May through August, the Midwest, northern Great Plains, and Great Lakes have the most tornadoes. By late fall, the highest frequency is again along the Gulf. Some hurricanes produce weak but numerous tornadoes in heavy rainsqualls at landfall. In most regions, tornadoes are most common during the warmest part of the day when surface air has heated to become most unstable. On the Plains and in the Midwest, most tornadoes occur during late afternoon to early evening. In the Gulf states, they can occur during day or night.

Waterspouts

A waterspout is a rapidly rotating column of air over water that has characteristics of a tornado. A waterspout always drops from a swelling cumulus or cumulonimbus cloud overhead. A *dark spot* on the water sometimes indicates a newly forming waterspout; in time a *funnel,* formed as atmospheric water condenses, drops from the cloud base. When a full waterspout forms, it connects the dark spot and the funnel. As wind speeds increase and the funnel descends, it may throw up a ring of *spray* as high as 100′ (30 m) above the surface of the water.

Waterspouts and Tornadoes

Most waterspouts are associated with routine swelling cumulus or cumulonimbus clouds that are not otherwise unusual. A very few strong waterspouts are tornadoes, spawned from mesocyclones, that form or move over water. Most, however, are less intense than tornadoes and normally lose their structure upon passing from water to land.

Season and Range

As water over 80°F (26°C) is very favorable for forming nontornadic waterspouts, they are mainly a warm-season event. In late summer, waterspouts frequently occur in the warm waters of the Florida Keys.

Dust Devils

A dust devil is a brief rotating column that reduces in size from the bottom upward until it dissipates in clear air. It has no connection to an overhead cloud. In the lowest portion near the ground, dust and debris spiral into the *core,* above which there is a more or less stable *vortex.* The top of the vortex usually diffuses into the air at around 100' to 300' (30–90 m); a few reach 2,000' (600 m). Dust devils can rotate clockwise or counterclockwise. Several can be in view at the same time. Most dust devils are harmless, but winds up to 90 mph (153 kph) have been recorded.

Dust Devils and Tornadoes Dust devils are not connected to cumulus clouds; their tops are in clear air. Much smaller and weaker than tornadoes, they arise from quite different atmospheric conditions: strong convection during sunny, hot, calm summer days. If a dust column reaches into a swelling cumulus cloud, it is a weak tornado raising dust.

Season and Range Dust devils occur most frequently over hot deserts with strong surface heating, though similar conditions sometimes occur on hot days elsewhere.

138

Hurricanes and Tropical Storms

Hurricanes have winds that exceed 74 mph (119 kph). *Tropical storms* have winds between 39 and 73 mph (63–117 kph). *Tropical depressions* have winds of less than 39 mph (63 kph). Hurricanes originate over warm tropical waters as depressions and then become tropical storms. Stronger storms have an *eye* (a calm center) and an *eye wall,* a bank of clouds surrounding the eye that contains the highest winds; it reaches from near the ocean to 40,000′ (12 km) high or more. Hurricanes are fed by *spiral bands* of swelling cumulus and cumulonimbus clouds that flow in from hundreds of miles away, often from the south and east. Winds are strong and gusty in the spiral bands.

As a hurricane approaches, there are lulls in the wind and rain until the strongest winds, in the eye wall, arrive, followed by calm conditions in the eye, then strong winds from the opposite direction. Convection in the spiral bands releases huge amounts of latent heat, which fuels the storm system. Above the spiral bands is the *outflow* of the anvil shield, which removes inflowing air that rises in the eye wall and is critical to the storm's maintenance.

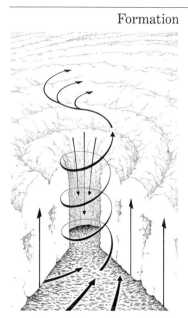

Formation

For a hurricane to form, the ocean surface must be at least 80°F (26°C). Tropical depressions start over warm oceans between 10°N and 30°N latitude. (South of 10° Earth's rotation is too weak to generate a rotating storm; north of 30° the water is usually too cool.) The most common starting feature is a tropical wave of low pressure moving east to west in the lower atmosphere. The critical stage of hurricane formation is the development of the eye wall. Continued inflow at lower levels and outflow at upper levels are also necessary.

Most of the U.S. is affected every year by tropical systems. Among the impacts are rising water levels—storm surges—of up to 10′ (3 m), especially to the right of the storm (relative to its forward motion), inundating low-lying shore areas. Tides also rise several feet, and waves may reach 25′ (8 m) or more. Winds from the eye wall and spiral bands can be severe. Some systems remain as tropical depressions or storms for many days, while others quickly intensify into hurricanes in less than 24 hours. Due to their size, hurricanes and tropical storms may take a day or more to completely pass over a given location.

Season and Range

The strongest tropical storms and hurricanes are most likely to occur in late summer along the Atlantic and Gulf of Mexico coasts. The normal tropical storm season lasts for six months, from the beginning of June to the end of November. The Gulf of Mexico and the Caribbean Sea spawn most of the early-season storms that affect the U.S.; many of these storms originate from stalled late-season cold fronts that penetrate unusually far south. June storms are generally small in size and weak in intensity. Activity increases slightly in July, and storms grow in size. A marked increase in both frequency and intensity takes place in August, and the place of origin expands eastward into the broad Atlantic Ocean. The height of the hurricane season is early September, when the tropical latitudes of the Atlantic Ocean, now warmed to 80°F (26°C) and up, become the principal storm-generating area. One-third of all North Atlantic storms occur in September; during this part of the hurricane season large developing storms tend to cross the Atlantic from west Africa to North America. The season goes into a steady decline in late October. The Atlantic produces an average of ten tropical storms each year, six of which reach hurricane strength.

Satellite Views of Hurricanes and Tropical Storms

Meteorological satellites have provided unprecedented data on the general organization, detailed structure, and motion of hurricanes, tropical storms, and depressions. Satellite images made every half hour over hurricane-prone regions have become invaluable for tracking and warning by the National Weather Service.

Satellite Monitoring

Most tropical storms in the North Atlantic, Caribbean, and Gulf of Mexico are born out of an easterly wave condition in which a trough of low pressure in the lower atmosphere travels west, often moving from west African waters across the Atlantic Ocean to American shores. The succession of waves in the easterlies can often be seen in large views of the Atlantic. Although there are few clues that forecast which waves will develop, imagery is watched continually for features that indicate exceptional waves.

Satellite views clearly show the circulation patterns associated with tropical storms. In low levels near the ocean are rainbands spiraling inward toward the eye. Sometimes an eye and eye wall are directly visible in the satellite view. At the top is the cirrus shield's overcast

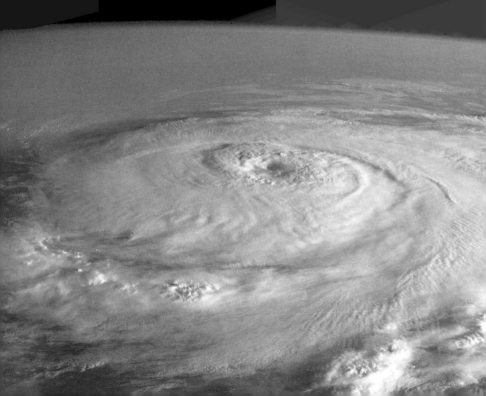

region, sometimes showing the outward spiral. In a well-developed storm, this shield hides many of the rainbands below. Such high clouds represent outward flow of air aloft that has been brought inward at low levels, ascended in the eye wall, and now moves outward in a continuation of the circulation. Any breakdown in these features causes a weakening of the entire storm system. Satellite monitoring pays particular attention to these movements.

Pacific Systems

Satellites have been especially important for new knowledge of tropical systems in the northeast Pacific off North America. Before routine satellite coverage, only a few systems were thought to occur there each year, but now the number is known to be two dozen or more per year. Tropical depressions, storms, and hurricanes are common off the west coast of tropical Mexico. Most move west-northwest over colder water, but some impact southern Arizona and California. While they rarely bring high winds to the U.S., traveling disturbances in the westerlies that are injected with mid-level moisture can be the source of deluges in arid and semiarid regions, and beneficial rains farther north and east. A few bypass the coldest Pacific water and reach Hawaii, as Hurricane Iniki did in 1992.

148

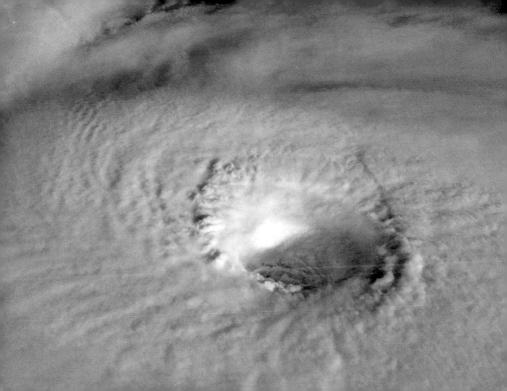

Snow

Snow consists of large and often complex ice crystals that originate in clouds. Snowflakes can be composed of single ice crystals or large, multi-crystal aggregates—as large as 2″ to 3″ (5–7.6 cm) across—which are seen especially in heavy snowfalls. The structure and size of the ice crystals depend on the temperature and moisture content of the air in which they form. Crystals that grow from the meager water supply at −20°F (−29°C) or below form hexagonal columns. At temperatures of about −10°F (−23°C) to 0°F (−18°C), most crystals are flat hexagonal plates. Large six-pointed dendrites form at 0° to 20°F (−18° to −7°C). Near-freezing temperatures bring splinter-shaped needles. Warm air contains more moisture than cold, allowing larger crystals to grow.

Most snow forms in supercooled water-droplet clouds, such as nimbostratus, swelling cumulus, and cumulonimbus, where water droplets turn to ice at temperatures colder than freezing. Under extremely cold conditions small ice crystals can grow directly from sublimation of water vapor, without the presence of any clouds. Snow can often be

seen falling from high clouds and evaporating long before reaching the ground. The clouds responsible for snow can result from fronts, frontal cyclones, orographic forcing, or convection. A *blizzard* exists when winds are at least 35 mph (56 kph), temperatures are below 20°F (−7°C), and there is considerable falling and blowing snow.

Effects   Most North Americans are affected during most winters by either direct snowfalls or their runoff flows. Snow can be a major problem when it falls in places where people are not accustomed to it. It has a major effect on our daily lives as a source of drinking water as well as water for forests, agriculture, and industry. The slow spring melting of snow in the mountains after a winter of heavy precipitation can make a significant contribution to a region's water supply. Snow also insulates the ground from cold and, in clouds, helps form other types of precipitation and lightning.

Avalanches are a common aftermath of excessive snow accumulation on mountain slopes, a situation often exacerbated by snow blowing over ridges to accumulate in cornices and unstable layers on the lee sides of ridges (downwind from the ridge). Many avalanches are triggered by collapsing cornices or by layers of snow

breaking free along their bases. A wet, cohesive snow layer can slide downhill as a *slab avalanche.* Looser snow usually disintegrates into a cloudlike *powder avalanche,* which can reach speeds of 200 mph (322 kph).

Season and Range

The only area of the mainland U.S. that has never had snow in recorded history is extreme southern Florida. Snow is infrequent in Mexico, except on its high volcanoes. Most of the rest of North America receives 1″ to 80″ (2.5–200 cm) of snow a year. Some typical seasonal snowfall totals are 8″ (20 cm) in central Oklahoma, 40″ (100 cm) in the Chicago area, and 80″ (200 cm) in Boulder, Colorado. The highest annual averages in North America are 500″ (1,270 cm) or more along the coastal mountains of Washington, British Columbia, and southeast Alaska. In the Rockies, snow exceeds 200″ (500 cm) a year on many higher slopes. East of the Rockies, highest seasonal totals reach 200″ in eastern Quebec and Newfoundland and in the snow belt east of Lake Ontario, and over 250″ (635 cm) on a few summits. In the high Arctic, annual snowfall averages only 20″ to 40″ (50–100 cm), mostly in spring and fall.

Ice

Ice is frozen water. The only two requirements for the formation of ice are the presence of water and subfreezing temperatures (below 32°F/0°C for fresh water; below 28°F/−2.2°C for seawater). Some varieties of ice, notably frost and black ice on highways, form when the air temperature is slightly above freezing but the ground temperature is below freezing. Evaporation on wet surfaces that are exposed to above-freezing air temperatures can lower the surface temperature below the freezing point. Water does not always freeze immediately when its temperature falls below 32°F (0°C). Water vapor needs a surface or a particle to condense on. At ground level, the air is full of minuscule condensation nuclei, such as particles of dust, pollen, sea salt, and smoke, that do not exist in undisturbed pure air, which can contain "supercooled" water drops down to −40°F (−40°C).

Frost Frost forms directly on the ground, leaves, rooftops, cars, windows, and other surfaces by sublimation of water vapor (a gas) directly to ice (a solid), bypassing the liquid phase.

For frost to occur, the surface it forms on must have a subfreezing temperature lower than the dew point (the temperature at which water vapor condenses), and the air temperature must be higher than the dew point.

Freezing Rain
Freezing rain occurs when water drops are formed aloft in a layer warmer than freezing. The precipitation falls to the ground and freezes on contact with objects that are colder than freezing. *Ice storms* coat objects with icy sheaths that can interrupt communications, transportation, and power distribution over large regions.

Sleet
Sleet occurs when raindrops are formed aloft in a layer of air that is warmer than freezing, and then descend through a deep layer of cold air near the ground, freezing into small globes or irregular grains of ice. The interior of an ice pellet may be part liquid. Sometimes both freezing rain and sleet occur during one storm period.

Hail
Balls or lumps of ice form when water drops are carried quickly to great heights, typically in a cumulonimbus, and then fall to the ground through a deep layer of moisture-saturated air. Hailstones usually are composed of layers of ice, the result of several stages of accretion as they pass

158

through successive moisture layers of the storm cloud. Hailstones can range from pea-size to grapefruit-size.

Ice on Bodies of Water

On small puddles and ponds, ice first freezes in a thin layer with a crystal structure. As the ice thickens, the crystal structure becomes less apparent, in part because of air bubbles trapped in the ice. On lakes large enough to have waves, the first ice to form is a thin surface layer of slush. Eventually the surface ice grows into small floes of pancake ice. If the lake is small enough or the temperatures stay cold enough, the floes may freeze together into a fairly solid sheet of pack ice. During cold and windy weather, spray may coat ships, offshore installations, and shoreline structures with massive loads of ice. Unlike most substances, water expands when it freezes, so that ice floats on top of a body of water. Without the insulating effect of floating ice sheets, surface water would lose heat more rapidly; some large bodies of water, such as the Arctic Ocean, Hudson Bay, and the Great Lakes, might eventually freeze up completely and remain frozen year-round, with a tremendous chilling effect on the climate.

River Floods

A river flood—the overflow of a watercourse beyond its natural banks or ordinary boundaries—may result from snowmelt in the spring or from ice jams during the winter or spring. Flooding is worse when storms or heavy precipitation add water to ground that is frozen, paved, or already saturated or that has an extremely dry, hard surface. The Mississippi River floods of 1993 are a prime example of river flooding that resulted from prolonged rainfall over a wide area. Tropical storms and hurricanes often drop enormous amounts of water into rivers and estuaries, adding to the height of storm-induced tidal surges. Flooding rivers take days, weeks, and sometimes months to build to the inundation level and return to normal.

Effects   Floods of all types take an average annual toll of 200 lives and \$2 billion of damage in the U.S. In 1972 Tropical Storm Agnes caused the flow of the Susquehanna River to approach that of the Amazon. In 1937 floods removed 300 million tons of topsoil in the Ohio Valley; not all was lost to the sea, but much was deposited downstream.

Flash Floods

Flash floods are sudden rises and falls of streams, usually resulting from brief but intense rainfalls over localized areas. They take tens of minutes to a few hours to make their impact and return to normal again. For example, in 1976 the Big Thompson Canyon (Colorado) flash flood killed 139 people; up to 10″ (25 cm) of rain fell from a stationary thunderstorm in less than two hours over a rugged canyon, resulting in a wall of water estimated at 19′ (6 m) high. Flash floods generally strike at night.

Causes   Streams and rivers confined to narrow valleys or steep canyons in mountains or hilly country are subject to sudden flooding when heavy rain falls in their watershed. Gentle terrain is vulnerable to flash floods after heavy rain in nearby mountains.

Effects   Flash floods are often lethal. Their sudden arrival and swiftly moving water can easily trap and drown unsuspecting victims. During the last few decades, flash flooding became the largest weather-related killer in the U.S. The best response to a flash flood is to get to higher ground immediately.

Drought

Drought is a sustained and abnormal lack of precipitation for a given region. What qualifies as a drought in a rainy location may represent adequate precipitation for another region. Most major droughts persist for months or years and cover thousands of square miles. Droughts are distinct from perpetual dryness in arid or semiarid climates and from normal, seasonal dry spells. Variation is normal in precipitation; some regions have more natural swings from dry to moist than others.

Causes   Most often the atmosphere has nearly the same moisture content during a drought as during moist periods. But during a drought, thunderstorms, tropical storms and hurricanes, or cold fronts and other traveling weather systems are weak or less frequent than usual.

Effects   Ski areas, reservoirs, agriculture, water supplies, river traffic, lawns, water tables, topsoil, permafrost depths, and wetlands are all affected by drought. Yields from entire categories of crops can decrease 10 to 20 percent or more during a widespread drought.

Dust Storms

In a dust storm, wind-borne dust or sand is lifted high enough above the ground (to eye level or higher) to reduce horizontal visibility to ⅝ mile (1 km) or less. Winds greater than 40 mph (64 kph) are needed to raise particles to such a level. Some dust storms result from high winds in large traveling weather systems. A dust storm that results from a thunderstorm outflow is called a *haboob* (a Sudanese name).

Effects
Dust storms can raise soil particles as high as 15,000′ (4,500 m) and hinder all types of travel. Once raised, the dust can travel long distances over a period of days. Dust from the Great Plains has been tracked with satellites into the Atlantic; dust from the Sahara reaches Florida in midsummer. A milky white or red hue around the midday Sun may indicate dust aloft.

Season and Range
Most U.S. dust storms occur in late winter and early spring in the Southwest and Plains states; others occur in the dry valleys of the West. Dust storms are more extensive and intense in periods of drought. Dust is easily raised when soil is fine and dry, and when there is little plant cover.

168

Fog

Essentially a water-droplet cloud with its base at the ground, fog reduces visibility below about $\frac{1}{2}$ mile (1 km). It occurs when the atmosphere at the ground holds as much water vapor as it can contain at that temperature. When temperatures are much below freezing, ice crystals may form in sufficient quantity to make *ice fog*.

Formation

Air holds more moisture at warm temperatures than cold. Fog often forms when moist air cools to its saturation point (dew point). Such situations include cooling at night, moist air flowing over a cold surface, or an upslope flow of air at a large elevation change. Fog also forms when winds, such as ones that blow from the ocean onto land, add moisture to the air until it reaches the saturation point at the ground. Fog often thins or disappears during the day as land warms and temperatures increase. Small cumulus or stratocumulus clouds may form as fog lifts.

Haze and Smog

Water vapor that causes less of a reduction in visibility than fog, haze can also occur before and after fog. Smog is formed when smoke mixes with fog; the combination of smoke and haze is also mistakenly described as smog.

170

Volcanic Clouds

Volcanoes explode into the troposphere, producing a wide variety of forms, including small puffs of smoke like cumulus clouds or a steady stream of ash similar to an altocumulus layer. A few exude massive amounts of rock and ash in a formation that acts as a giant cumulonimbus, complete with tornadoes and lightning. Volcanic clouds are darker than water-vapor clouds.

Effects   Volcanic material can reduce temperatures in the immediate area by around 20°F (11°C). The eruption of Mount Pinatubo in the Philippines in 1991 cooled much of the world by as much as a degree for several years. Volcanic ash near eruptions has been responsible for several near-crashes of jet aircraft in the U.S. and Canada just after the formation of undetected volcanic clouds that were embedded in heavy cloud cover. As it becomes distributed over large areas, volcanic ash can enrich the soil.

Season and Range   The existence of volcanic clouds is totally dependent on volcanic activity, which has no seasonal variations. In North America active volcanoes occur from Alaska southward to California and in central Mexico.

Volcanic Twilight

When volcanoes eject sulfur dioxide gas above 60,000′ (18,000 m), it may combine with water vapor to form a heavy layer of sulfuric acid droplets that can linger for months or years. This high layer is illuminated by the Sun long after sunset and before sunrise, causing changing colors in the sky. Purple and lavender hues at twilight are usually associated with a significant amount of volcanic material aloft (left photo). Normal twilight colors tend toward yellow, green, and blue (right photo). The 1991 eruption of Mount Pinatubo in the Philippines emitted the largest volume of small particles into the stratosphere in this century. The resulting cloud spread over much of the world within a year.

Effects   Volcanic activity occurs somewhere on Earth every day, but only events that eject large amounts of fine particles above the troposphere change the colors of twilight. The magnitude of an eruption does not necessarily predict the amount of material that may remain aloft. However, the brightness of volcanic twilights indicates the thickness of the stratospheric particle layer.

174

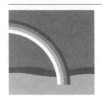

Auroras

Auroras occur when high-energy particles from solar flares reach Earth and are deflected by its magnetic fields toward the polar regions. The particles strike molecules and atoms in our ionosphere, exciting their electrons and causing them to glow in spectacular displays of light. This colorful phenomenon is called the aurora borealis (or northern lights) in the Northern Hemisphere, and the *aurora australis* in the Southern Hemisphere. Auroral displays occur in arcs, bands, rays, or curtains that wave and flicker. The predominant color is the green glow of oxygen atoms; at higher altitudes and lower latitudes oxygen may glow deep ruby-red. At lower altitudes, nitrogen molecules emit blue and red colors on the bottom fringes of arcs and rays. The enormous electron flow during auroras can cause rapid fluctuations in Earth's magnetic field that affect power transmission and communications.

Season and Range

From Alaska east through Labrador to Norway the northern lights occur 100 or more nights a year. They extend across southern Canada about 30 nights yearly and dip once a year into the southern U.S.

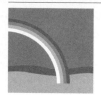

Halos and Parhelia

When ice-crystal clouds are uniform across the sky, halos—whitish or colored rings and arcs—sometimes appear around the Sun or Moon. The inner edge of a halo is weak red and the outer edge is weak blue, but the color may be mostly white. Often only part of a halo is visible. The most common type appears at an angle 22° from an imaginary line drawn between the observer and the Sun or Moon. The halo is produced by the refraction of light that enters one face of a six-sided ice crystal and leaves by a second face. In order for a full 22° halo to be seen, the sky must be filled with hexagonal ice crystals falling randomly, which frequently occurs. The rarer 46° halo is a prismatically colored circle with an angular radius of 46° from the viewer.

Parhelia A parhelion (also called a *mock sun* or a *sun dog*) is a weakly colored bright spot that occurs to the right and/or left of the Sun. It is also produced by the refraction of light through hexagonal ice crystals. Parhelia follow the elevation of the Sun above the horizon. Similar bright spots near the Moon are known as *paraselenae* (or *moon dogs*).

178

Coronas

A corona is a nearly round area of pastel-hued clouds in the general direction of the Sun or Moon. A corona is more diffuse and not in a perfect circle, as is a halo. Coronas near the Sun are often uncomfortably bright; those around the Moon are easier to observe.

Formation

Coronas are created by the diffraction of light waves passing through relatively thin clouds containing small water droplets of a fairly uniform size. The characteristic droplet sizes and concentrations for the creation of a corona are found most often in altocumulus and altostratus clouds. Lenticular altocumulus clouds and pileus clouds (small horizontal clouds) atop cumulonimbus are also sources of iridescence and coronas.

Appearance

Small droplets within the cloud layer produce larger ring patterns than do larger droplets. Distinct colors and nearly circular rings result from nearly uniform droplet sizes, while muted colors and irregular ring patterns are produced by droplets of varying sizes. A perfectly uniform cloud layer with droplets of uniform size produces alternating bluish and reddish rings centered on the Sun or Moon.

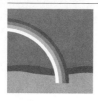

Iridescence

Iridescence is a colorful patch or band in clouds some distance from the Sun or Moon. Iridescence is not centered on the light source, as is a halo or corona, and it usually does not appear circular. When very near the Sun, the colors may be uncomfortably bright.

Formation

Iridescence is produced by diffraction of light through water droplets of a fairly uniform size. (Diffraction is the breakdown of the color spectrum due to changes in light as it travels around the edges of water droplets.)

Appearance

Iridescence is often seen along the fringe on the edge of a cloud layer. Colors are pale shades of neon blue, orange, purple, and pink. Iridescence occurs in relatively thin altostratus, altocumulus, and cirrocumulus clouds. It can also be seen in the thin, smooth pileus cloud (a small horizontal cloud) draped on top of a cumulonimbus tower.

Crepuscular Rays

Crepuscular rays are alternating bright and dark rays that appear to radiate from the Sun's location at a time when the Sun itself is hidden by clouds either below or above the horizon. The scattering of sunlight by haze, dust, or mist in the air enhances the contrast between rays.

Formation

The scattering of sunlight by molecules and particles in the air makes the bright rays visible. Scattering occurs when small particles diffuse a portion of the incoming radiation in all directions. (The blue color of the sky results from scattering of the Sun's light by air particles.)

Appearance

Crepuscular rays appear to radiate upward from below the horizon after sunset or before sunrise. The rays are nearly parallel, although they appear to diverge. Rays may also radiate from the Sun when it is behind a cloud at any time of day. Typically, crepuscular rays need a scattered or broken cloud cover or the presence of one or more cumulus or cumulonimbus clouds. But sometimes a single ray will shine through the only cloud-free patch in the sky. A mountain range may produce crepuscular rays at sunrise or sunset even when there are no clouds in the sky.

Rainbows

A rainbow is a multicolored arc or circle centered on the point in the sky directly opposite the Sun. Rainbows need two features: falling rain and sunlight shining on the rain. They chiefly occur in association with isolated rainshowers and thunderstorms.

Formation   Rainbows are formed by the reflection of light within raindrops. Some of the light entering a drop is refracted into its component colors, reflected off the back inside wall of the drop, and refracted again as it exits the drop. Spherical drops smaller than $\frac{1}{16}''$ (1.6 mm) result in the best rainbows. Sunlight reflecting twice inside the raindrops produces a larger but fainter secondary rainbow. Moonbows and fogbows also form by the same processes.

Season and Range   All year in the tropics and subtropics; chiefly summer elsewhere. Over most of North America, rainshowers are more numerous in the afternoon than in the morning; thus most rainbows are seen in the east as the Sun sinks in the west. Rainbows are most frequent in climates where showers tend to be isolated and least frequent in consistently wet and cloudy climates.

186

A Summary of Cloud Types and Their Significance

Cirrocumulus Most often found with larger-scale weather systems. A small or isolated cloud insignificant. Several patches or a widespread layer may indicate increasing moisture or instability ahead of an advancing weather system.

Cirrostratus If becoming more dense and increasing, watch for an increase in the number of other clouds over a period of several hours; more active weather may be approaching. Also may be part of an inactive high-level moisture outflow far from the source, such as from the southwest U.S. monsoon or from tops of subtropical or tropical disturbances.

Contrails A few isolated contrails are of no significance. If the number of these and other stratiform clouds are both increasing, a large-scale weather system may be approaching.

Altocumulus Large areas of altocumulus usually accompany active, moving weather systems and the flow of significant moisture over hundreds of miles. When they last only a few hours or cover only a small portion of the sky, there is probably not an organized weather system in the area. If clouds are also at other levels, a more organized system is at, near, or approaching the area. When the amount and thickness of altocumulus are increasing and changes are approaching from the wind direction at cloud level, moister weather is likely aloft and perhaps at the ground.

Altostratus If increasing in coverage and approaching from the wind direction at cloud level, widespread precipitation may be expected. It may not occur directly in the area, however, since altostratus clouds are often carried hundreds of miles without much change.

Stratus Indicate saturation near the ground. They can form as a result of the cooling of Earth's surface. May also come from the flow of moist, cold air into a region at low altitudes. May also represent transition to or from fog and cumulus clouds.

Stratocumulus May be a recurring local condition along a coast at certain times of the year. In other locations, stratocumulus clouds accompany larger-scale, traveling weather systems, typically after the passage of a cold front or other disturbance, and especially in the afternoons. Often seen before the sky clears. May be the final result of decrease in cloud cover before a new weather cycle begins.

Nimbostratus Important to watch, as they always bring precipitation. If cloud is convective, a thunderstorm may develop quickly.

Small Cumulus A sign of generally good weather if they do not grow much. The formation of shallow clouds within a blue sky may be only a temporary stage, as some continue to grow into larger cumulus. When continuous upward currents of warm air are present, small cumulus can become swelling cumulus. When small cumulus occur in the morning throughout midday, the atmosphere may be unstable enough for them to grow into larger clouds later in the day. If small cumulus are the only convective clouds during the afternoon, there is no threat of showers in the short term. In some areas and situations, small cumulus may begin to grow at or after sunset and develop into large nighttime thunderstorms. Small cumulus clouds can coexist along with thunderstorms and cumulonimbus in other portions of the sky.

Swelling Cumulus (Cumulus Congestus) Should be watched over time to see whether they are the largest cumulus that will occur, or if this is just a temporary stage on the way to cumulonimbus. The earlier in the day that congestus clouds form, the more significant they are as indicators of strong storms later that day. If there are quite a few swelling cumulus clouds on a sunny summer day in the morning to early afternoon, the atmosphere may be unstable enough for them to reach the cumulonimbus stage, which can produce significant rain or severe weather. If cumulus congestus are the largest cumulus clouds that have appeared by late afternoon, they probably will not grow any larger for the rest of that day. Other factors

may help them grow further just before or after sunset. In tropical weather, cumulus congestus may produce momentary heavy rain showers, or form into cloud lines that last for hours.

Cumulonimbus Very important to monitor whenever there is a good supply of low-level moisture and strong upward forcing of the air. The earlier in the day that a cumulonimbus forms, or the more organized and vigorous it appears, the more likely it is to become significant. When cumulonimbus is arranged in rows or complexes, the likelihood of severe weather increases. Such weather can include tornadoes, waterspouts, and funnel clouds; brief to prolonged heavy rain, hail, sleet, snow, and flash floods; strong winds and turbulence at any level of the atmosphere, including gust fronts, squall lines, and microbursts; and lightning.

Mammatus When mammatus are weak, in the distance, or not approaching, the parent storm is not likely to be very important. If those conditions are not met, the sky

should be watched for possible severe weather. Mammatus clouds most often follow the most active growth stage of a cumulonimbus within minutes to an hour or more.

Gust Front Wind typically increases in speed and changes direction when a gust front passes, creating an aviation hazard. Winds greater than 100 mph (160 kph) over large areas are typically from *squall lines* —gust fronts organized together. Winds may increase up to 15 minutes after the passage of a gust front.

Mountain-induced Cumulus These clouds contain areas of strong winds and the chance of increasing wind speed inside the cloud and sometimes on the ground; they require caution in the operation of aircraft. Thunderstorms may grow later in the day if conditions are suitable for further growth to the cumulonimbus stage.

Mountain-induced Stratiform Strong winds and turbulence can exist inside the clouds or on the ground below. Winds as high as 100 mph (160 kph) or more in these clouds are a potential hazard for aircraft.

Index

Credits

Photographers

Gary Braasch (93)
Bill Bunting (119, 121)
Ed Cooper (171)

DEMBINSKY PHOTO ASSOCIATES:
Stan Osolinski (39, 49)
Rod Planck (105)

David Hoadley (81, 85)
Ronald L. Holle (45, 47, 53, 57, 65, 79, 89, 99, 175 left)
Richard Keen (43, 55, 59, 83)
Martin Kleinsgorge (75)
Chlaus Lotscher (173)
Alan Moller (115, 117, 123, 125, 131, 185)
Courtesy NASA (31, 147, 149)
National Hurricane Center (33)
National Oceanic Atmospheric Administration (35, 37)
Paul Neiman (113, 187)
Scott Norquay (137)
Pekka Parviainen (183)

PHOTO/NATS: Mary Clay (159)

PHOTO RESEARCHERS:
Stephen J. Krasemann (91, 157)
Carl Purcell (143)
Pekka Parviainen/Science Photo Library (71)

John Eastcott & Yva Momatiuk (163)
Jack Finch/Science Photo Library (177)

Sylvia Schlender (161)
Science Source (29)
Scott T. Smith (51, 63, 103, 107, 151, 179)
Arjen & Jerrine Verkaik/SKYART (41, 69, 73, 77, 87, 95, 97, 101, 109, 129, 169, 175 right, 181)
Voscar/The Maine Photographer (155)
Steve Warble (67)

WEATHERSTOCK: W. Balzer (135)
Keith Brewster (127)
Ed Darack (153)
Warren Faidley (111, 133, 139, 141, 145, 165)
Richard Lewis (167)

Fred Whitehead (61)

Cover Photograph: Positive flash lightning by Warren Faidley/WEATHERSTOCK
Title Page: Thunderstorm with hail by David Hoadley
Pages 26–27: Cumulonimbus calvus, detached anvil, by Arjen & Jerrine Verkaik/SKYART

Staff

Prepared and produced by
Chanticleer Press, Inc.

Founding Publisher: Paul Steiner
Publisher: Andrew Stewart

Staff for this book:

Managing Editor: Edie Locke
Art Director: Amanda Wilson
Production Manager: Susan
Schoenfeld
Photo Editor: Giema Tsakuginow
Photo Assistant: Consuelo
Tiffany Lee
Publishing Assistant: Alicia Mills
Project Editor: Amy K. Hughes
Text Editor: Patricia Fogarty
Natural Science Consultant:
Richard Keen
Picture Editor: Alexandra Truitt
Picture Researcher: Jerry Marshall
Illustrations: Acme Designs, Ed Lam
Original series design by
Massimo Vignelli

All editorial inquiries should be
addressed to:
Chanticleer Press
665 Broadway, Suite 1001
New York, NY 10012

To purchase this book, or other
National Audubon Society
illustrated nature books, please
contact:
Alfred A. Knopf, Inc.
201 East 50th Street
New York, NY 10022
(800) 733-3000